Holt Mathematics

Chapter 7 Resource Book

HOLT, RINEHART AND WINSTON
A Harcourt Education Company
Orlando • **Austin** • New York • San Diego • London

Printed in the United States of America

ISBN 0-03-078397-6

7 8 9 10 170 10 09 08

CONTENTS

Blackline Masters

Date__________

Dear Family,

In this chapter, your child will learn about the fundamentals of plane geometry, including classifying and naming geometric figures, finding angle measures, and identifying polygons in the coordinate plane. Your child will also learn about transformations and symmetry of plane figures. This work will provide your child with the foundation needed for further study of geometry and assist in preparation for careers in a variety of areas such as graphic arts, fashion, and architectural design.

Your child will learn to name **points, lines, planes, segments,** and **rays.**

Look at the figure.

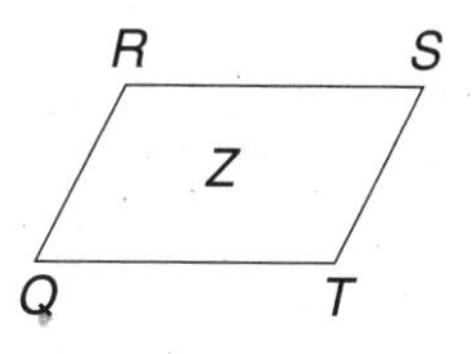

1. Name four points.

point *Q*, point *R*, point *S*, point *T*

2. Name four lines.

$\overleftrightarrow{QR}$, $\overleftrightarrow{RS}$, $\overleftrightarrow{ST}$, $\overleftrightarrow{TQ}$

3. Name a plane.

plane *Z* or plane *QRS*

Any three points in the plane that form a triangle can be used.

4. Name four segments.

$\overline{QR}$, $\overline{RS}$, $\overline{ST}$, and $\overline{TQ}$

5. Name five rays.

$\overrightarrow{QR}$, $\overrightarrow{RS}$, $\overrightarrow{ST}$, $\overrightarrow{TQ}$, $\overrightarrow{RT}$

Your child will learn to identify **parallel** and **perpendicular lines** and the **angles** formed by a **transversal.** A transversal is a line that intersects any two or more other lines.

Measure the angles formed by the parallel lines and the transversal. Which angles seem to be congruent?

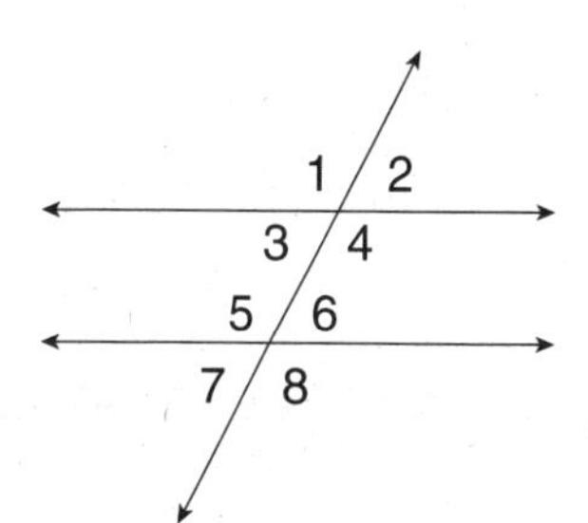

Angles 1, 4, 5, and 8 all measure **130°.**
Angles 2, 3, 6, and 7 all measure **50°.**

Your child will also learn how to identify the different types of triangles and how to find the angles of triangles.

The three angle measures of a triangle add up to 180°.

Find *x* in the acute triangle.

$$62° + 33° + x° = 180°$$
$$95° + x° = 180°$$
$$-95° \quad -95°$$
$$x° = 85°$$

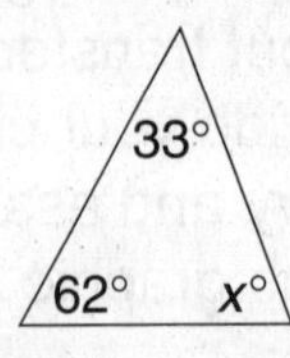

Your child will also learn about transforming plane figures using translations, rotations, and reflections. A **translation** slides a figure along a line without turning. A **rotation** turns the figure around a point, called the center of rotation. A **reflection** flips the figure across a line to create a mirror image.

Identify each as a translation, rotation, reflection, or none of these.

A

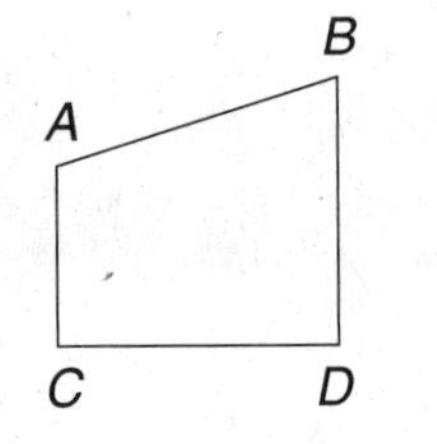

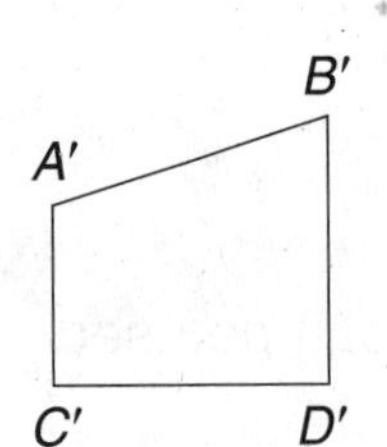

translation

B

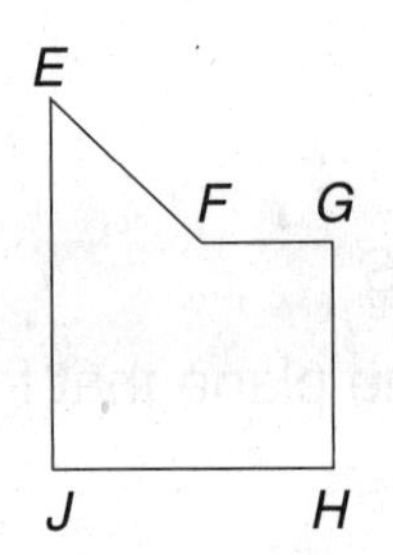

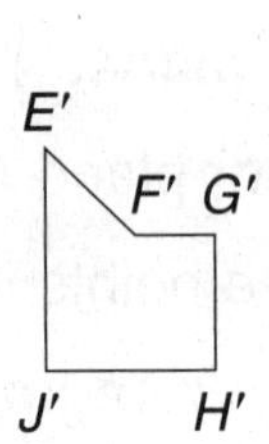

none of these

C

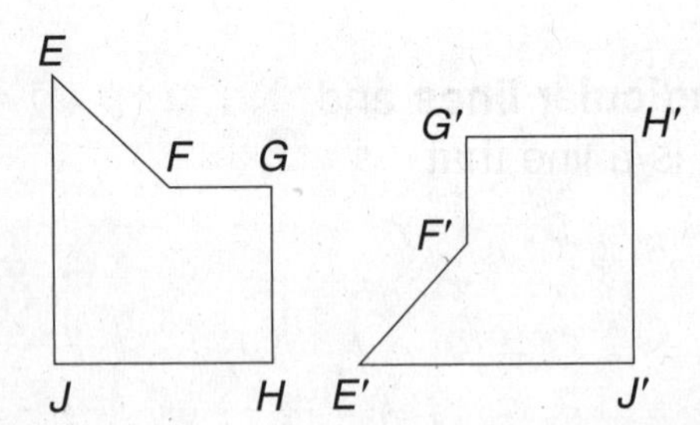

rotation

D

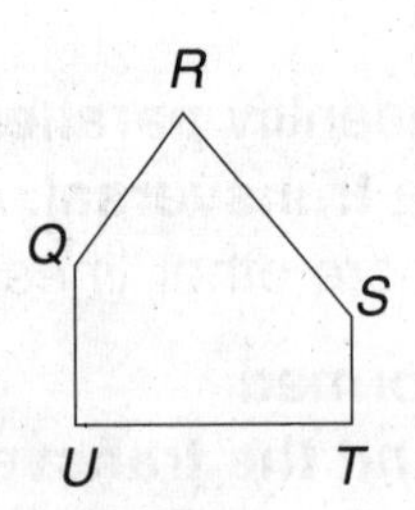

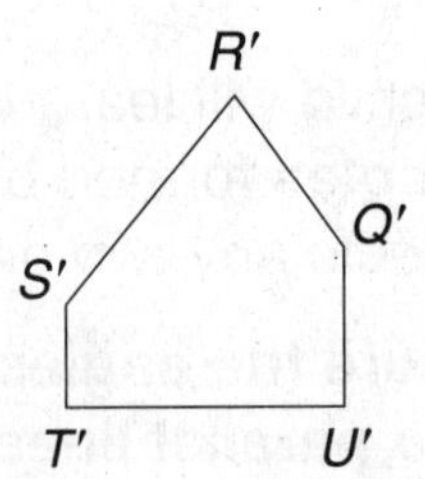

reflection

For additional resources, visit go.hrw.com and enter the keyword MT7 Parent.

Name ______________________ Date __________ Class __________

LESSON 7-1

Practice A

Points, Lines, Planes, and Angles

Use the diagram to name each figure.

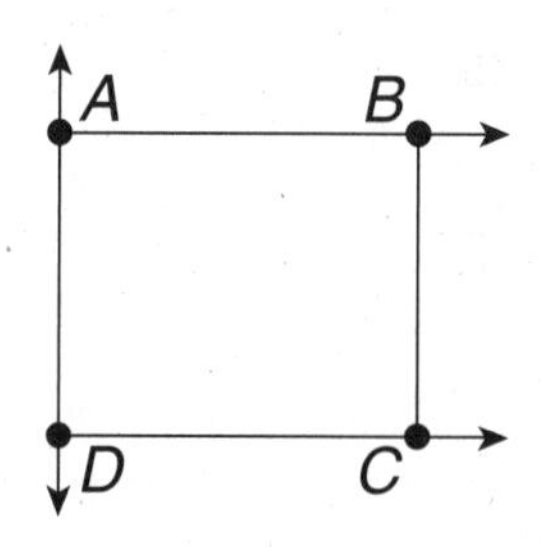

1. four points

____

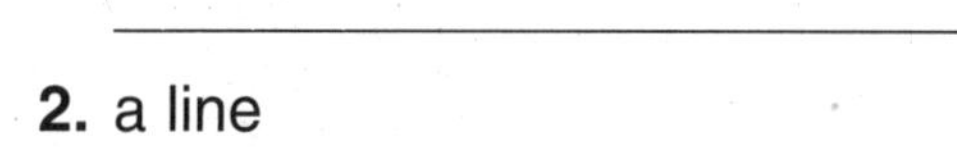

2. a line

____

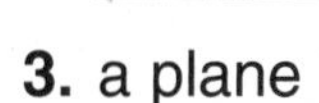

3. a plane

____

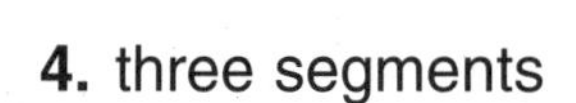

4. three segments

____

5. four rays

____

Use the diagram to name each figure.

6. a right angle

____

7. two acute angles

____

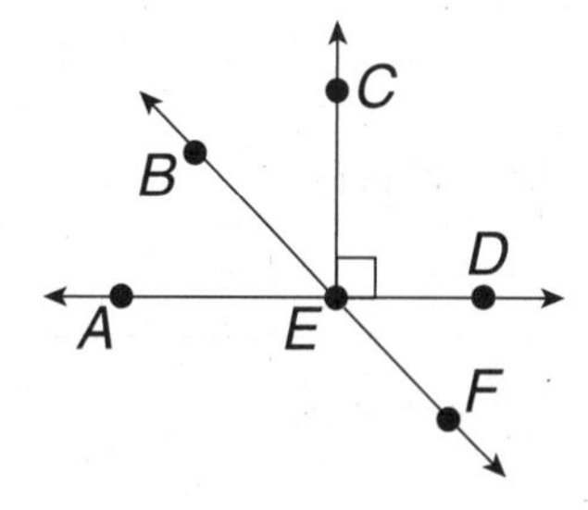

8. two obtuse angles

____

____

9. pair of complementary angles

____

____

10. two pairs of supplementary angles

____

____

In the figure, $\angle 1$ and $\angle 3$ are vertical angles, and $\angle 2$ and $\angle 4$ are vertical angles.

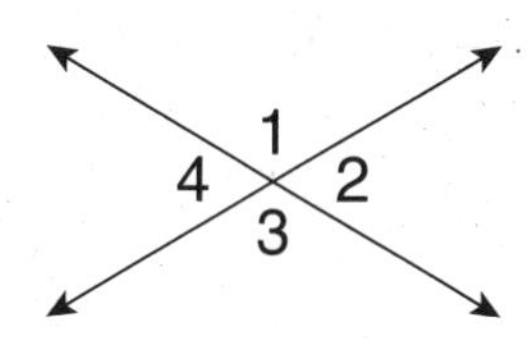

11. If $m\angle 1 = 120°$, find $m\angle 3$.

____

12. If $m\angle 2 = x°$, find $m\angle 4$.

____

Name ______________________ Date __________ Class __________

LESSON 7-1

Practice B

Points, Lines, Planes, and Angles

Use the diagram to name each figure.

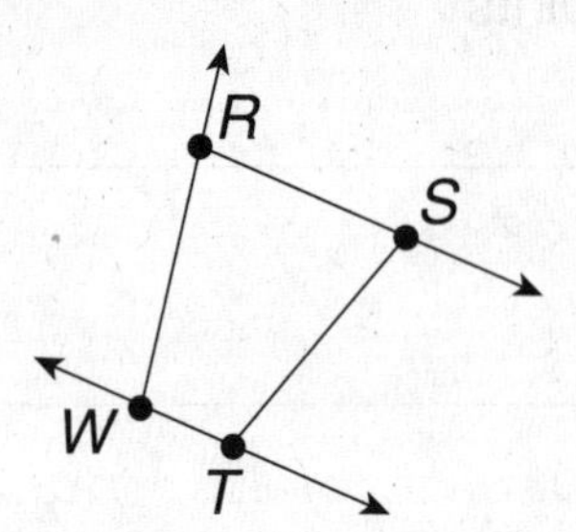

1. four points

____

2. a line

____

3. a plane

____

4. three segments

____

5. four rays

____

Use the diagram to name each figure.

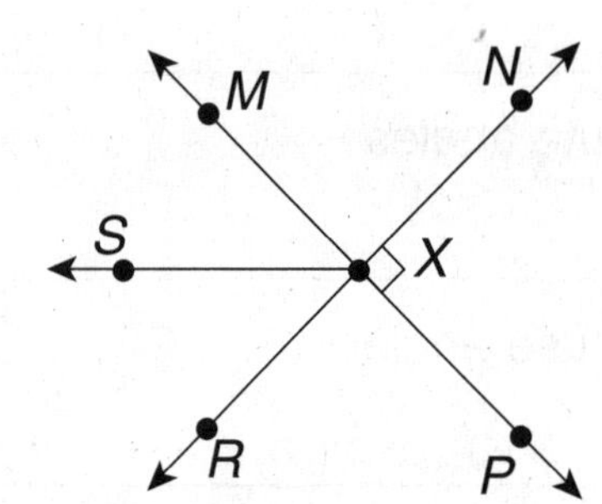

6. a right angle

____

7. two acute angles

____

8. two obtuse angles

____

9. a pair of complementary angles

____

____

____

10. three pairs of supplementary angles

____

____

____

In the figure, $\angle 1$ and $\angle 3$ are vertical angles, and $\angle 2$ and $\angle 4$ are vertical angles.

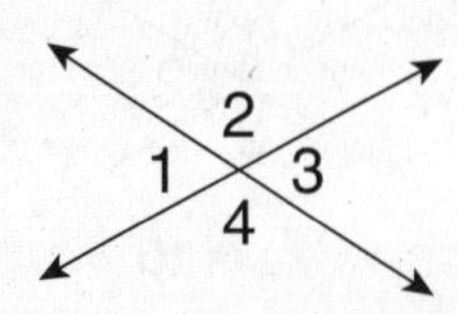

11. If $m\angle 2 = 110°$, find $m\angle 4$.

____

12. If $m\angle 1 = n°$, find $m\angle 3$.

____

Name ______________________________ Date ____________ Class ____________

LESSON **7-1**

Practice C

Points, Lines, Planes, and Angles

Use the diagram to solve.

Write *true* or *false*. If a statement is false, rewrite it so it is true.

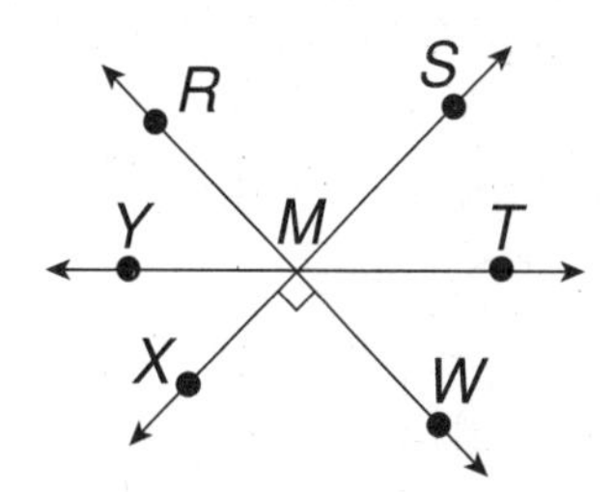

1. RW is a line.

____

2. Rays YM and MS make up line YT.

____

____

3. Angle RMT is an obtuse angle.

____

4. $\angle WMY$ and $\angle RMY$ are supplementary angles.

____

5. $\angle YMX$ and $\angle SMT$ are supplementary angles.

____

____

6. $\angle XMY$ and $\angle YMR$ are complementary angles.

____

____

7. If $m\angle SMT = 48°$, then $m\angle TMW = 48°$.

____

____

8. If $m\angle SMY = 135°$, then $m\angle RMT = 135°$.

____

____

9. If $m\angle TMW = x°$, then $m\angle RMT = 180° - x°$.

____

10. $m\angle RMY + m\angle SMT + m\angle RMS = 180°$.

____

In the figure, $\angle 1$ and $\angle 3$ are vertical angles.

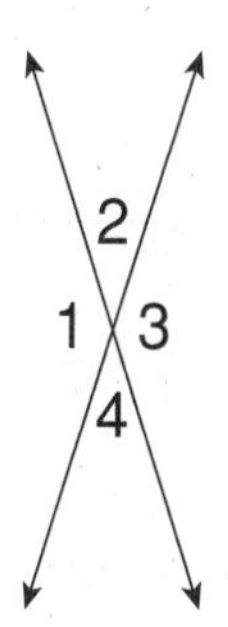

11. If $m\angle 1 = 123°$, find $m\angle 3$.

____

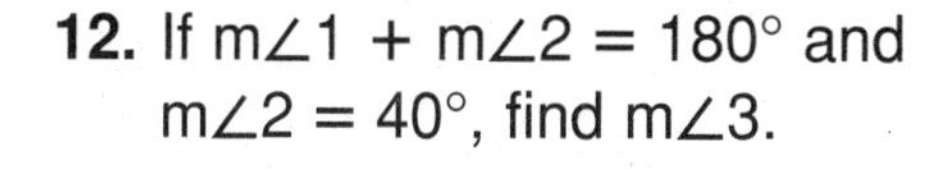

12. If $m\angle 1 + m\angle 2 = 180°$ and $m\angle 2 = 40°$, find $m\angle 3$.

____

Name ______________________ Date __________ Class __________

LESSON 7-1

Reteach

Points, Lines, Planes, and Angles

Figure	Description	Diagram	Notation: Write	Notation: Read
Line	an infinite collection of points with no beginning and no end	ℓ, A, B	$\overleftrightarrow{AB}$ or $\overleftrightarrow{BA}$ or ℓ	line *AB*, line *BA*, line ℓ
Line Segment	part of a line, with two endpoints	A, B	$\overline{AB}$ or $\overline{BA}$	line segement *AB* line segment *BA*
Ray	part of a line, with one endpoint	A, B	$\overrightarrow{AB}$	ray *AB*

Use the diagram, to name each type of figure.

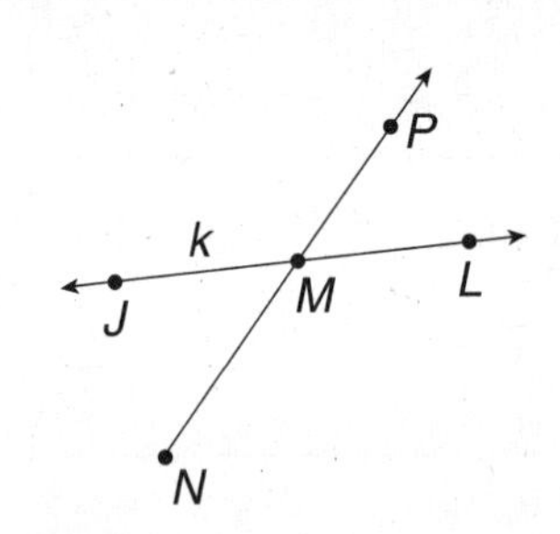

1. $\overrightarrow{MP}$ ______________________

2. k ______________________

3. $\overline{MN}$ ______________________

4. $\overline{LJ}$ ______________________

5. $\overleftrightarrow{JL}$ ______________________

Acute Angle	Right Angle	Obtuse Angle	Straight Angle
Measures between 0° and 90°	Measures exactly 90°	Measures between 90° and 180°	Measures exactly 180°.

Use the diagram to name each type of angle.

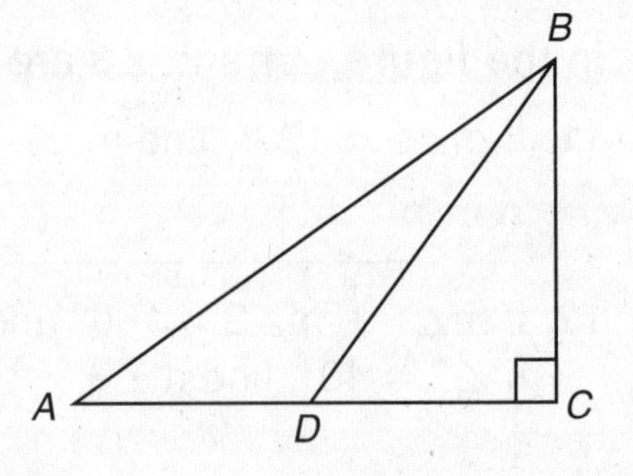

6. $\angle BCD$ ______________________

7. $\angle BAD$ ______________________

8. $\angle BDA$ ______________________

9. $\angle CDA$ ______________________

10. $\angle BDC$ ______________________

11. $\angle ABC$ ______________________

Name ______________________ Date __________ Class __________

LESSON 7-1

Reteach

Points, Lines, Planes, and Angles (continued)

Complementary Angles	Supplementary Angles	Vertical Angles
Two angles whose measures have a sum of 90°.	Two angles whose measures have a sum of 180°.	Intersecting lines form two pairs of vertical angles.
$\angle A$ and $\angle B$ are complementary angles.	$\angle C$ and $\angle D$ are supplementary angles.	$\angle a$ and $\angle b$, $\angle c$ and $\angle d$ are pairs of vertical angles.

Use the diagram to complete.

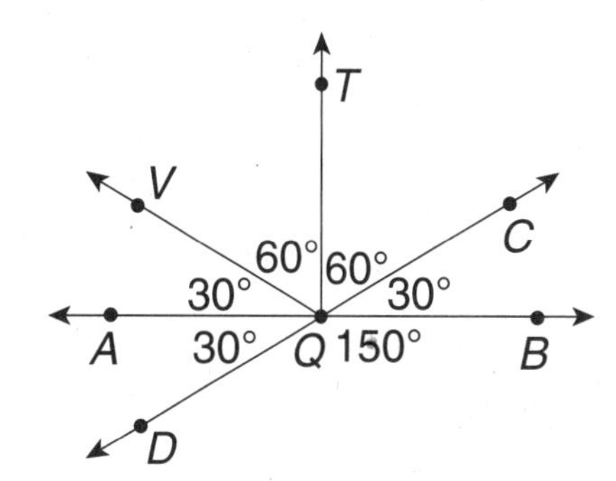

12. Since $\angle AQC$ and $\angle DQB$ are formed by intersecting lines, $\overleftrightarrow{AQB}$ and $\overleftrightarrow{CQD}$, they are:

____

13. The sum of the measures of $\angle AQV$ and $\angle VQT$ is: ______
So, these angles are:

____

14. The sum of the measures of $\angle AQC$ and $\angle CQB$ is: ______

So, these angles are: ______________________

Congruent figures have the same size and shape.
The symbol ≅ means *is congruent to.*

Complete these statements about congruence.

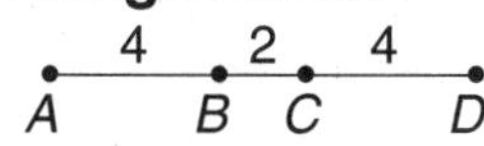

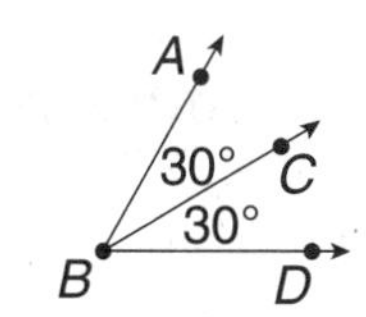

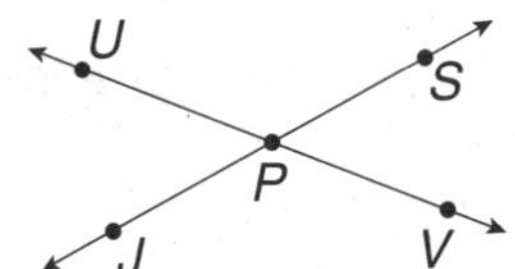

15. Since $AC = 6$ units and $BD = 6$ units,

then $\overline{AC}$ ______ $\overline{BD}$.

16. Since $m\angle ABC = 30°$ and $m\angle CBD = 30°$

then $\angle ABC$ ________.

17. Since vertical angles are congruent,

then $\angle UPJ$ ________.

Name ______________________________ Date ____________ Class ____________

LESSON 7-1

Challenge

Let's Meet!

Materials needed: paper strips, index cards, and scissors

1. Use a flat surface such as the top of your desk to represent a plane. Use strips of paper to represent lines. Move the lines around in the plane (**coplanar lines**) to determine the number of intersections that are possible. Summarize your results in a table.

Number of Coplanar Lines	Possible Number of Points of Intersection
2	0 or 1
3	
4	
5	

2. Slit one index card and connect two cards to model two intersecting planes.

a. What is the intersection of two planes?

____

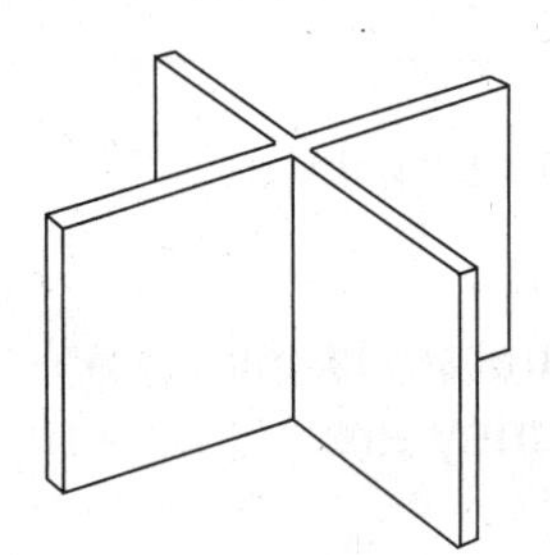

b. Mark the diagram to illustrate the intersection of the two planes.

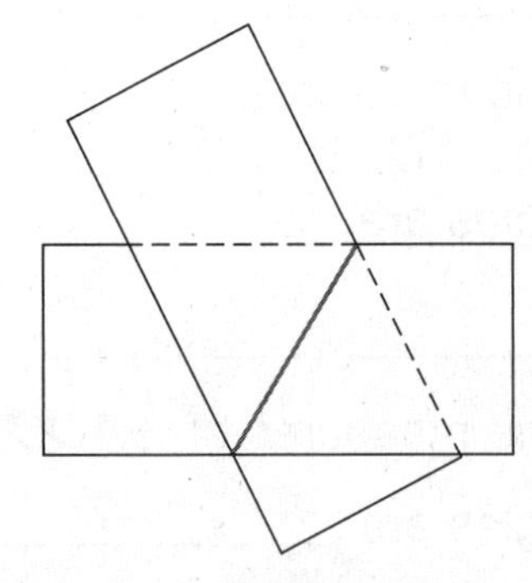

3. Using index cards to represent planes, determine the number of intersections that are possible. Summarize your results in a table.

Number of Planes	Possible Number of Lines of Intersection
2	0 or 1
3	
4	
5	

Name ______________________________________ Date ___________ Class ___________

LESSON **7-1**

Problem Solving

Points, Lines, Planes, and Angles

Use the flag of the Bahamas to solve the problems.

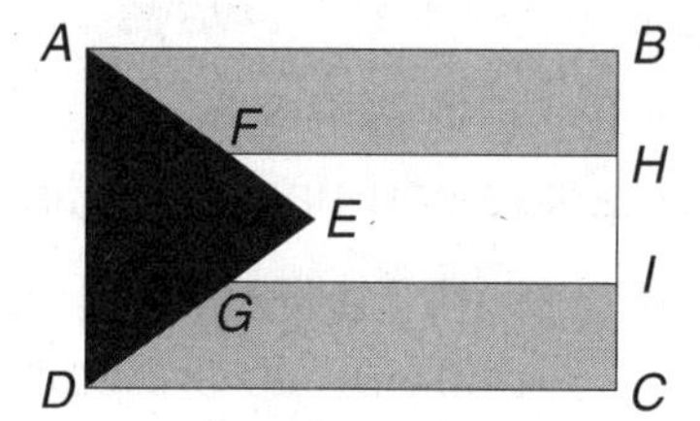

1. Name four points in the flag.

____

2. Name four segments in the flag.

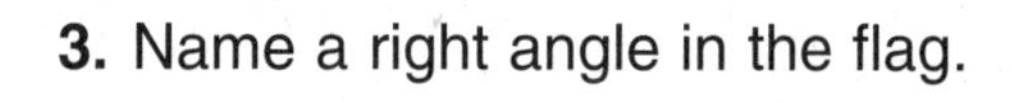

3. Name a right angle in the flag.

____

4. Name two acute angles in the flag.

____

5. Name a pair of complementary angles in the flag.

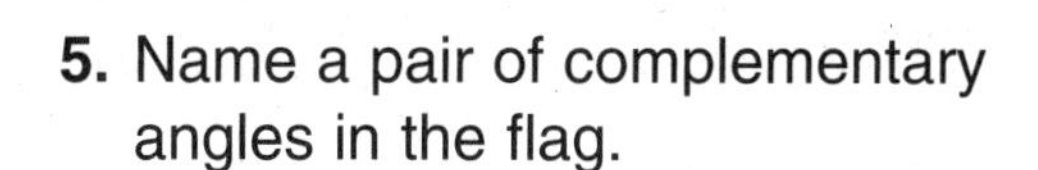

6. Name a pair of supplementary angles in the flag.

____

The diagram illustrates a ray of light being reflected off a mirror. The angle of incidence is congruent to the angle of reflection. Choose the letter for the best answer.

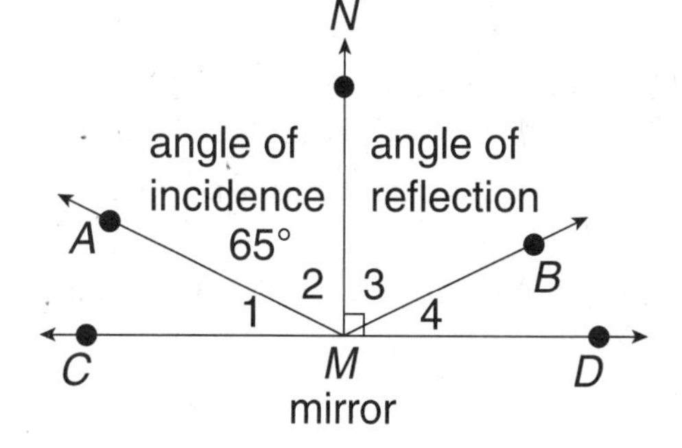

7. Name two rays in the diagram.

A $\overline{AM}$, $\overline{MB}$ **C** $\overrightarrow{MA}$, $\overrightarrow{MB}$

B $\overrightarrow{MA}$, $\overrightarrow{BM}$ **D** $\overrightarrow{MA}$, $\overrightarrow{MB}$

8. Name a pair of complementary angles.

F $\angle NMB$, $\angle BMD$ **H** $\angle CMA$, $\angle AMD$

G $\angle AMN$, $\angle NMB$ **J** $\angle CMA$, $\angle DMB$

9. Which angle is congruent to $\angle 2$?

A $\angle 1$ **C** $\angle 3$

B $\angle 4$ **D** none

10. Find the measure of $\angle 4$.

F 65° **H** 25°

G 35° **J** 90°

11. Find the measure of $\angle 1$.

A 65° **C** 25°

B 35° **D** 90°

12. Find the measure of $\angle 3$.

F 90° **H** 35°

G 45° **J** 65°

Name ______________________ Date ____________ Class ____________

LESSON 7-1

Reading Strategies
Understanding Vocabulary

A **line** is a straight path that extends forever in both directions. A line can be named in two different ways.

- A lowercase letter can name a line.
- Two points on a line can name the line.

A **line segment** is part of a line running between two endpoints. A line segment is named by its endpoints, using capital letters.

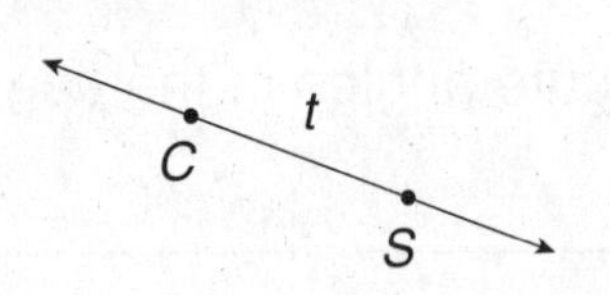

Line t or $\overleftrightarrow{CS}$ or $\overleftrightarrow{SC}$
Read: line *t*, or line *CS*, or line *SC*

$\overline{CS}$ or $\overline{SC}$
Read: segment *CS* or segment *SC*

Use this figure to answer each question.

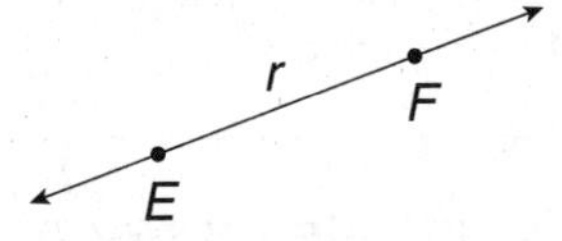

1. How would you name the line segment?

____

2. How would you name the line in this figure?

____

A **plane** is a flat surface that extends without end in all directions. You use three points that are not on a line to name a plane. Planes are named using capital letters. This plane could be named plane *MVW*.

3. What is another way to name this plane?

____

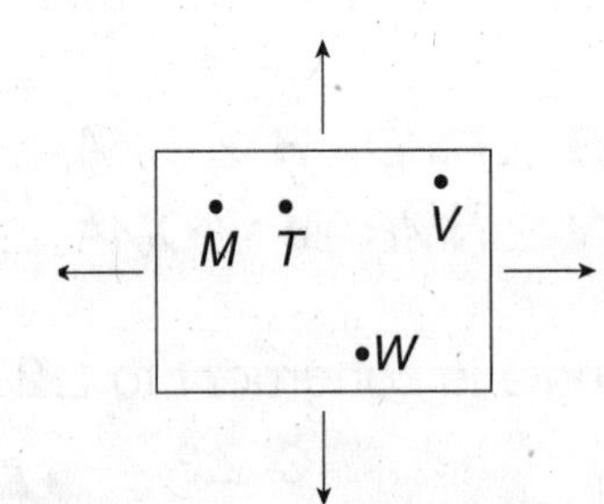

4. Draw a plane and identify three points in the plane that are not on a line.

____

____

Name ______________________ Date __________ Class __________

LESSON 7-1

Puzzles, Twisters & Teasers

What's Your Angle?

Circle the words below in the word search (horizontally, vertically or diagonally). Then find a word in the word search that answers the riddle. Circle it and write it on the line.

point	supplementary	plane	segment	ray
angle	complementary	line	congruent	vertical

```
C O M P L E M E N T A R Y S
O M H O Y U P U I O N W E U
N O L I T Y L T R Y G R T P
G I U N R T A W E R L U I P
R A Y T Q W N P L M E Y H L
U W D V G T E Q W E R T Y E
E E D C V F R T G B R H Y M
N O I U Y T R F V B A J K E
T I S E G M E N T U Y T F N
I M L P O K N B J L U H V T
M W D C F R T G H N I J U A
Q V E R T I C A L D C N H R
E P A N T S U J M K I O E Y
```

Why did the fireman wear red suspenders?

To keep his ___ ___ ___ ___ ___ on.

Name ______________________ Date ____________ Class ____________

LESSON **7-2**

Practice A

Parallel and Perpendicular Lines

1. Measure the angles formed by the transversal and the parallel lines. Which angles seem to be congruent?

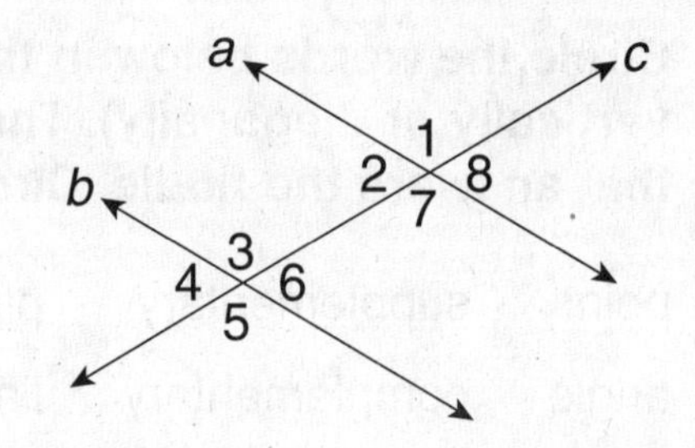

____

____

In the figure, line *r* ∥ line *s*. Find the measure of each angle.

r
x
s
130°
1
2
3
5
4
6
7

2. ∠1 ______ 3. ∠5 ______ 4. ∠6 ______

5. ∠7 ______ 6. ∠4 ______ 7. ∠3 ______

In the figure, line *m* ∥ line *n*. Find the measure of each angle.

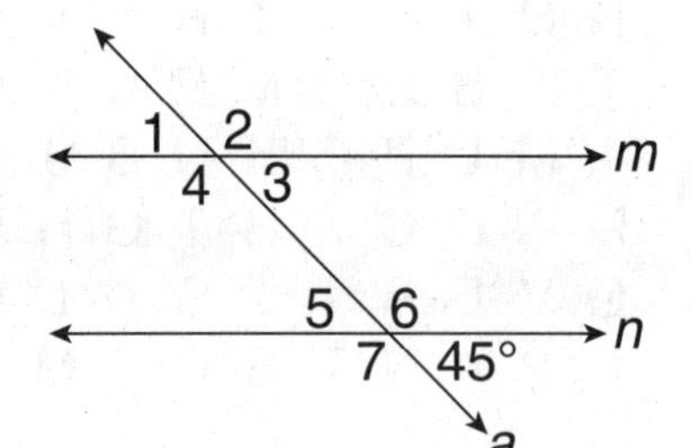

8. ∠1 ______ 9. ∠2 ______ 10. ∠3 ______

11. ∠5 ______ 12. ∠6 ______ 13. ∠7 ______

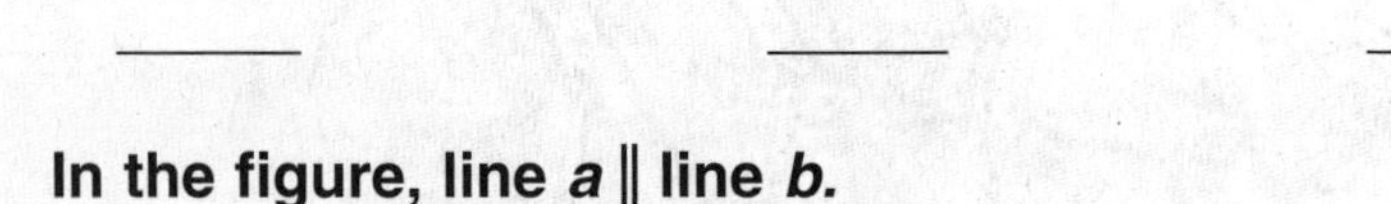

In the figure, line *a* ∥ line *b*.

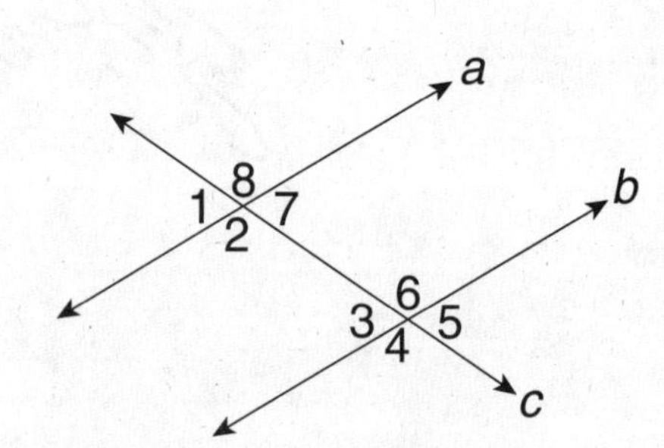

14. Name all angles congruent to ∠1.

____

15. Name all angles congruent to ∠2.

____

16. Name three pairs of supplementary angles.

____

17. Which line is the transversal?

____

Name ______________________ Date __________ Class __________

LESSON **7-2**

Practice B

Parallel and Perpendicular Lines

1. Measure the angles formed by the transversal and the parallel lines. Which angles seem to be congruent?

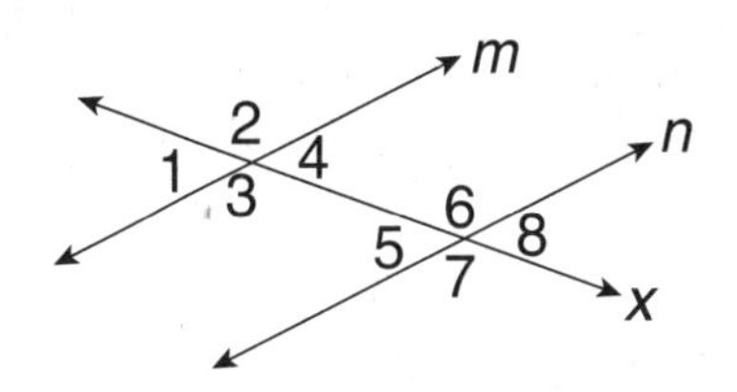

____

____

In the figure, line $m \parallel$ line n. Find the measure of each angle.

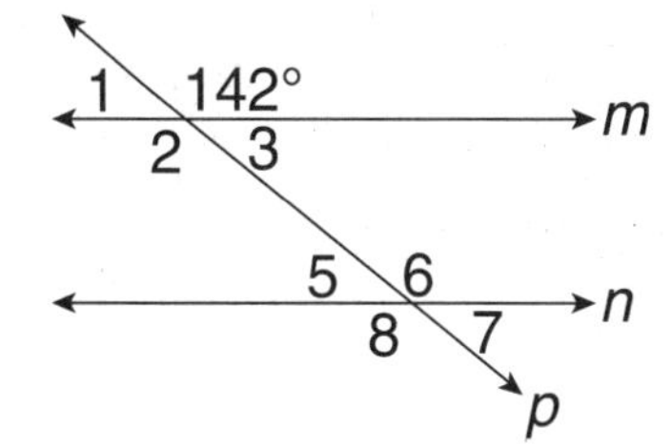

2. $\angle 1$ ______ **3.** $\angle 2$ ______ **4.** $\angle 5$ ______

5. $\angle 6$ ______ **6.** $\angle 8$ ______ **7.** $\angle 7$ ______

In the figure, line $a \parallel$ line b. Find the measure of each angle.

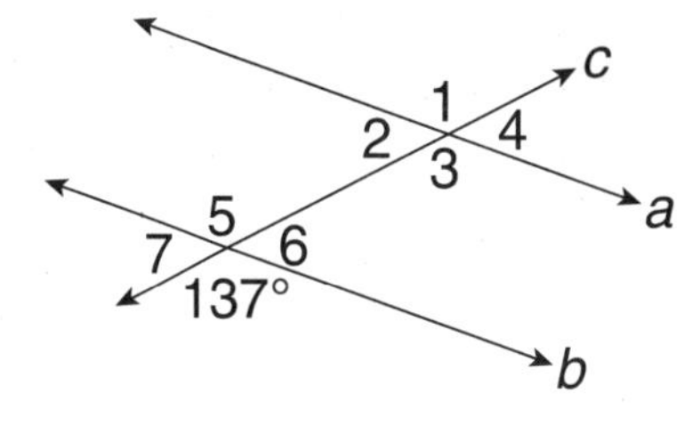

8. $\angle 2$ ______ **9.** $\angle 5$ ______ **10.** $\angle 6$ ______

11. $\angle 7$ ______ **12.** $\angle 4$ ______ **13.** $\angle 3$ ______

In the figure, line $r \parallel$ line s.

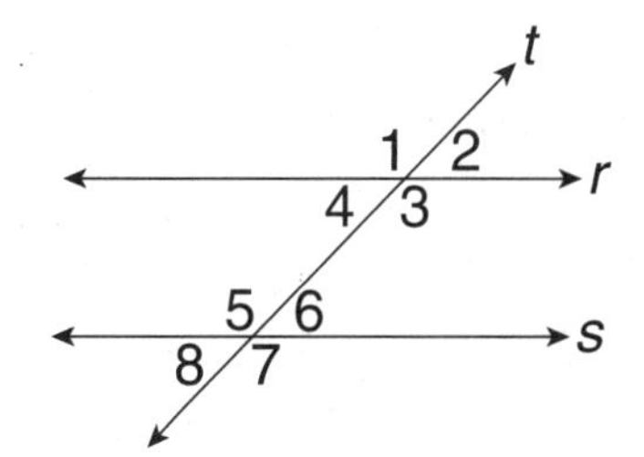

14. Name all angles congruent to $\angle 2$.

____

15. Name all angles congruent to $\angle 7$.

____

16. Name three pairs of supplementary angles.

____

17. Which line is the transversal?

____

Name ______________________ Date ____________ Class ____________

LESSON **7-2**

Practice C

Parallel and Perpendicular Lines

In the figure, line *a* ∥ line *b*.

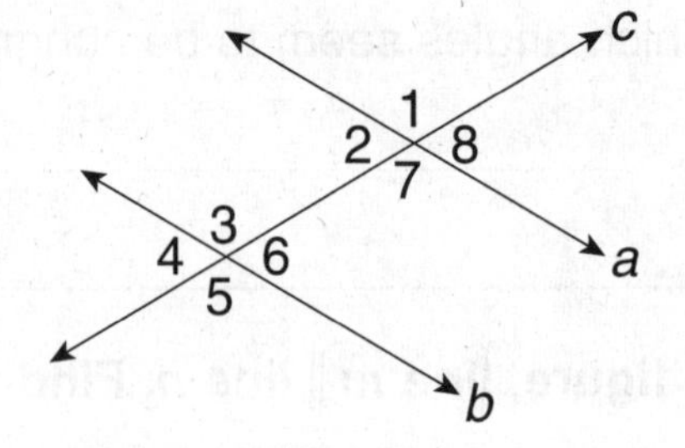

1. Name all angles congruent to ∠1.

____

2. Name all angles congruent to ∠2.

____

3. Name three pairs of supplementary angles.

____

4. Which line is the transversal?

____

5. If m∠7 is 131°, what is the m∠8?

____

6. If m∠4 is 57°, what is the m∠5?

____

7. If m∠3 is 127°, what is the m∠8?

____

8. If a transversal were drawn perpendicular to line *a* and line *b*, what would be the measure of the angles formed?

____

Draw a diagram to illustrate each of the following.

9. line *x* ∥ line *y* ∥ line *z* and transversal *k*

10. line *r* ∥ line *s* and transversal *m* with eight congruent angles

11. line *a* ∥ line *b* and transversal *t* with ∠1 ≅ ∠2 and ∠3 ≅ ∠4

Name ______________________ Date __________ Class __________

LESSON 7-2

Reteach

Parallel and Perpendicular Lines

Parallel Lines

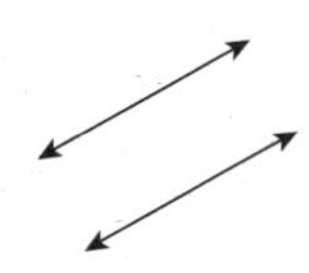

Parallel lines never meet.

Perpendicular Lines

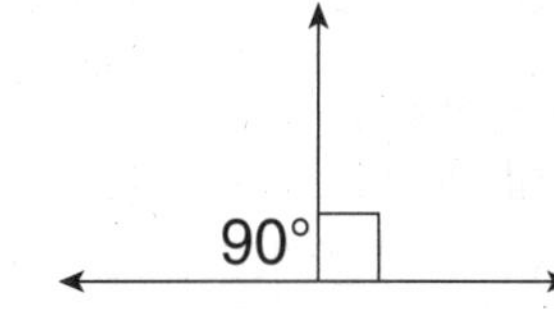

Perpendicular lines form right angles.

When parallel lines are cut by a **transversal,** 8 angles are formed, 4 acute and 4 obtuse.

The acute angles are all congruent.

The obtuse angles are all congruent.

Any acute angle is supplementary to any obtuse angle.

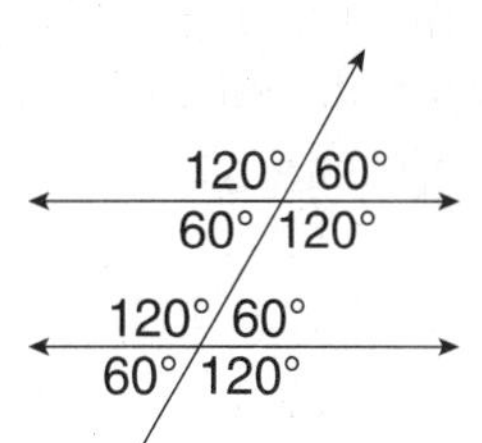

In each diagram, parallel lines are cut by a transversal. Name the angles that are congruent to the indicated angle.

1.

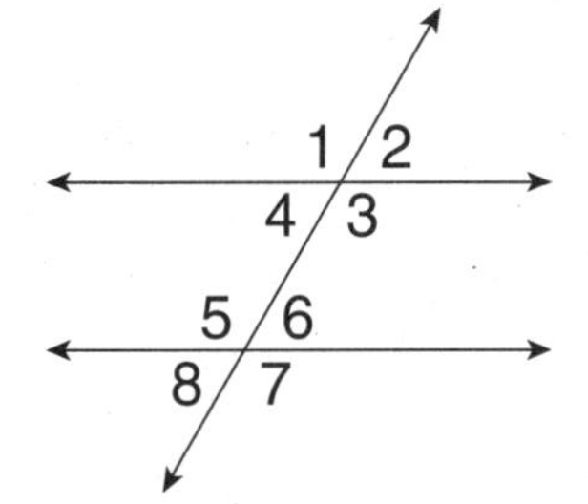

The angles congruent to $\angle 1$ are:

____

2.

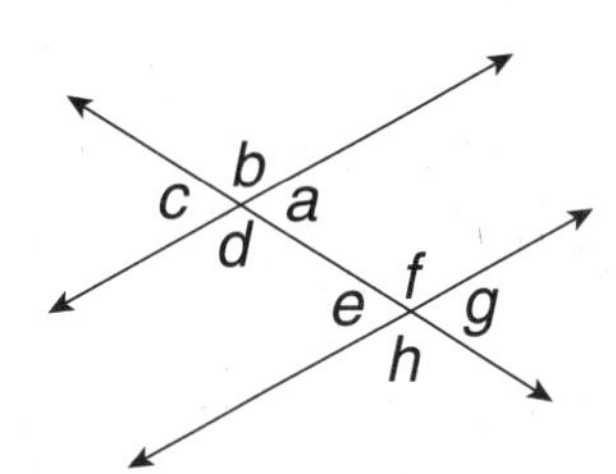

The angles congruent to $\angle a$ are:

____

3.

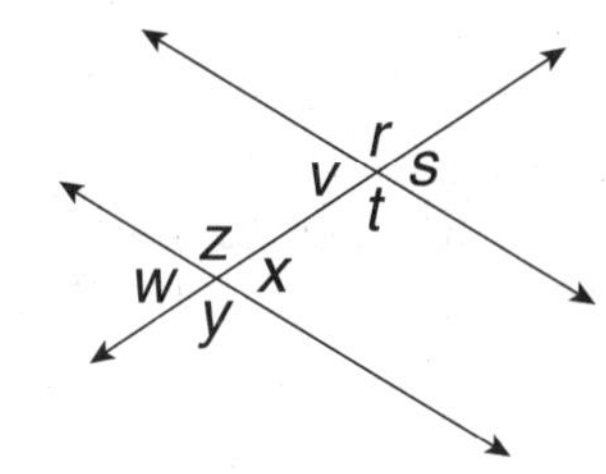

The angles congruent to $\angle z$ are:

____

In each diagram, parallel lines are cut by a transversal and the measure of one angle is given. Write the measures of the remaining angles on the diagram.

4.

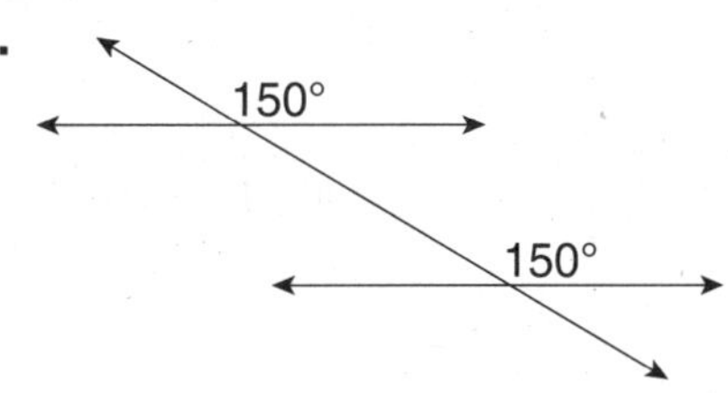

5.

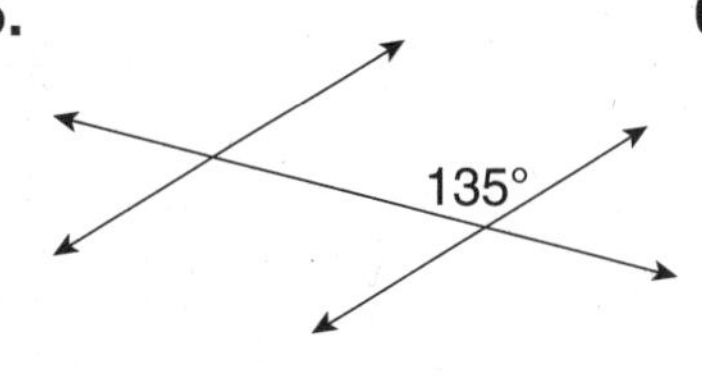

6.

25°

Name ______________________ Date ____________ Class ____________

LESSON **7-2**

Challenge
Pairing Off

When two parallel lines are cut by a transversal, eight angles are formed. Of these, four angles are between the parallel lines, **interior angles.**

1. In this diagram, name the four interior angles formed by the parallel lines and the transversal.

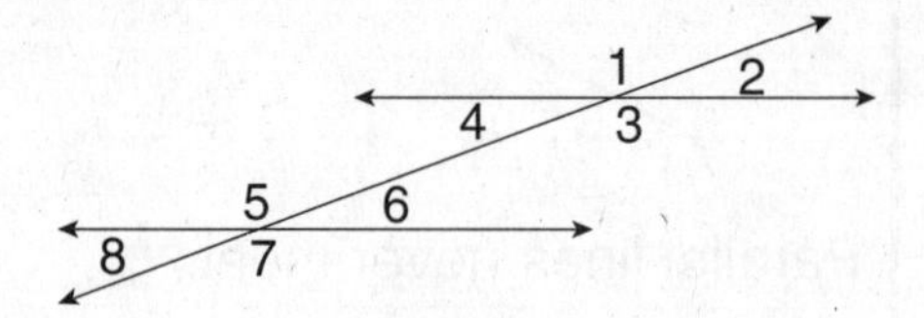

____

2. Think of the interior angles in pairs. Name the two pairs of interior angles that are on opposite sides of the transversal.

____

3. What is true about the measures of ∠3 and ∠5 in the diagram above? in the diagram at the right? Use a protractor to verify your conjecture.

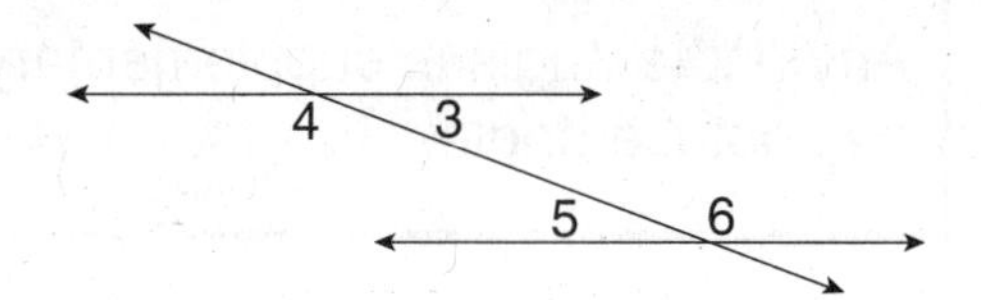

____

4. What is true about the measures of ∠4 and ∠6 in the diagram above? in the diagram at the right? Use a protractor to verify your conjecture.

____

5. Interior angles that are on opposite sides of the transversal are called **alternate interior angles.**

Draw a conclusion about the measures of alternate interior angles formed by parallel lines and a transversal.

____

____

Use your observation about the measures of alternate interior angles of parallel lines to find the measure of ∠*x* in each of these diagrams.

6.

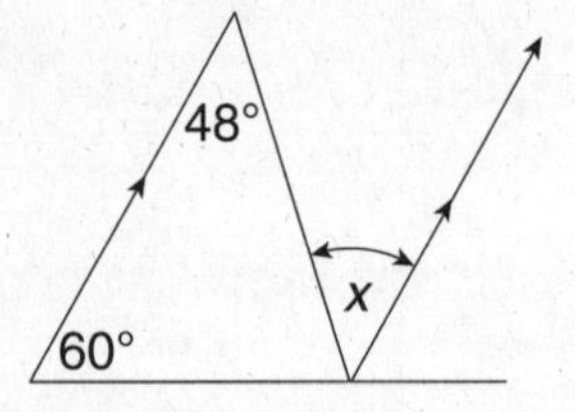

m∠*x* = ____________

7.

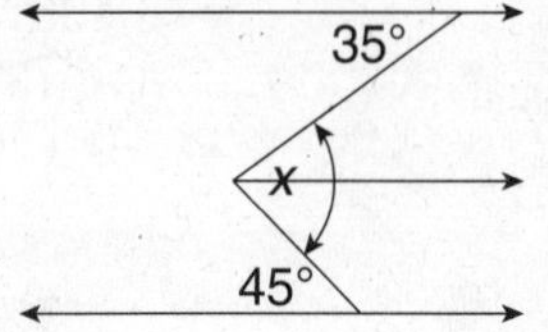

m∠*x* = ____________

8.

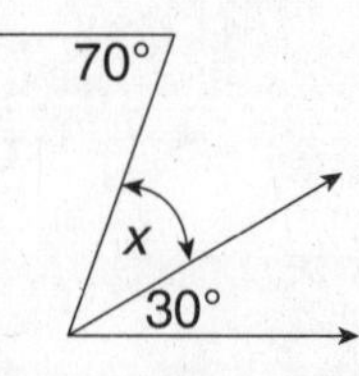

m∠*x* = ____________

Name ______________________ Date __________ Class __________

LESSON 7-2

Problem Solving

Parallel and Perpendicular Lines

The figure shows the layout of parking spaces in a parking lot. $\overline{AB} \parallel \overline{CD} \parallel \overline{EF}$

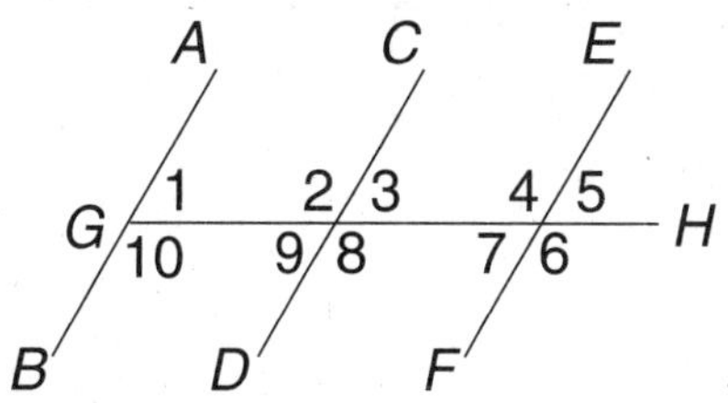

1. Name all angles congruent to ∠1.

____

2. Name all angles congruent to ∠2.

____

3. Name a pair of supplementary angles.

____

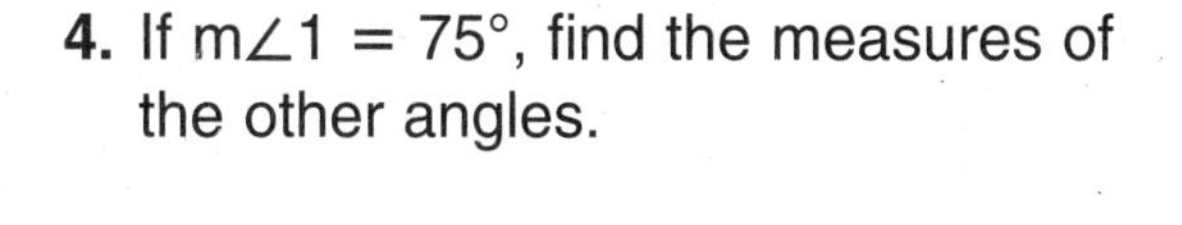

4. If m∠1 = 75°, find the measures of the other angles.

____

____

____

5. Name a pair of vertical angles.

____

6. If m∠1 = 90°, then $\overline{GH}$ is perpendicular to

______________________.

The figure shows a board that will be cut along parallel segments *GB* and *CF*. $\overline{AD} \parallel \overline{HE}$. Choose the letter for the best answer.

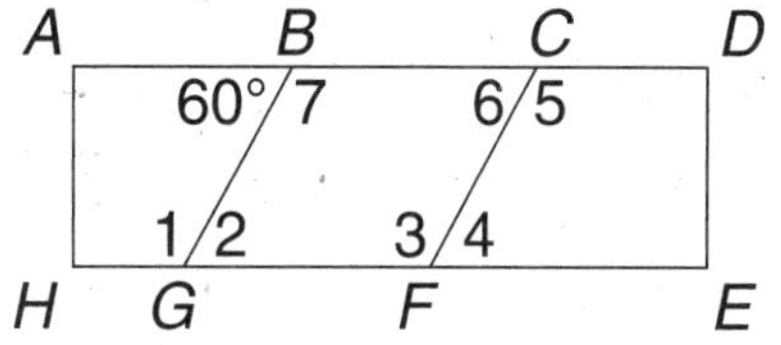

7. Find the measure of ∠1.

A 45° **C** 60°
B 120° **D** 90°

8. Find the measure of ∠2.

F 30° **H** 60°
G 120° **J** 90°

9. Find the measure of ∠3.

A 30° **C** 60°
B 120° **D** 90°

10. Find the measure of ∠4.

F 45° **H** 60°
G 120° **J** 90°

11. Find the measure of ∠5.

A 30° **C** 60°
B 120° **D** 90°

12. Find the measure of ∠6.

F 30° **H** 60°
G 120° **J** 90°

13. Find the measure of ∠7.

A 45° **C** 60°
B 120° **D** 90°

Name ____________________ Date __________ Class __________

LESSON 7-2

Reading Strategies

Understanding Symbols

The symbol || stands for **parallel.** Parallel lines are the same distance apart and never meet.

The yard lines marked on a football field are similar to parallel line segments.

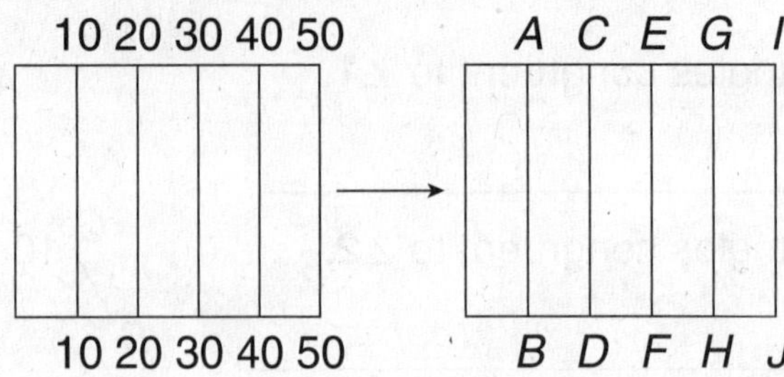

Segment *AB* is parallel to segment *CD*. This can be written: $\overline{AB} \parallel \overline{CD}$.

1. Identify another pair of parallel line segments in the figure above.

____

2. Use the || symbol to write how the line segments are related.

____

The symbol ⊥ stands for **perpendicular.** Perpendicular lines meet to form a square corner.

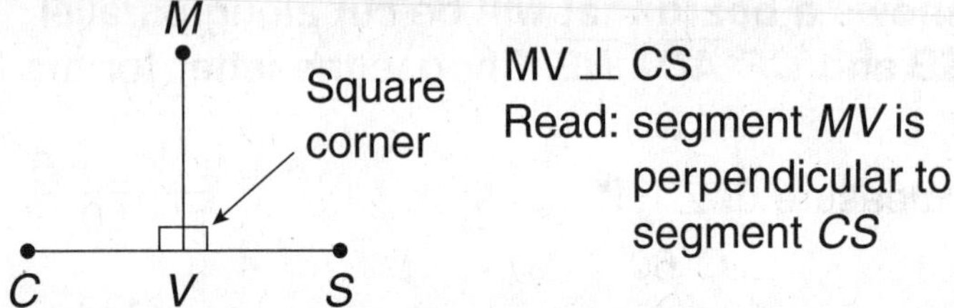

MV ⊥ CS

Read: segment *MV* is perpendicular to segment *CS*

A square in a corner is used as a symbol in a figure to show that lines are perpendicular.

Use this figure for Exercises 3–5.

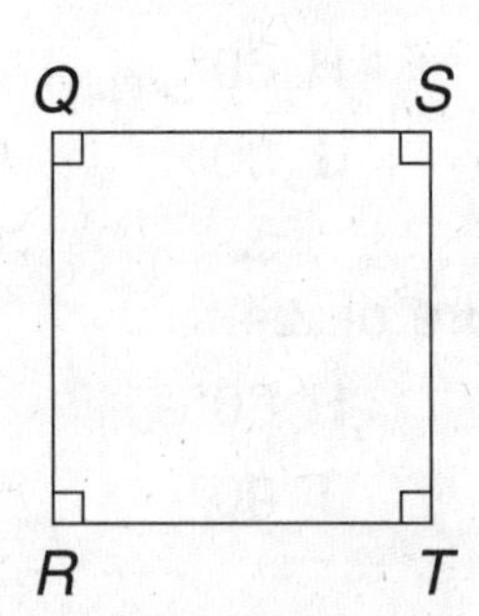

3. Segment *QR* is perpendicular to segment *TR*. Write this with symbols.

____

4. Identify two other line segments that are perpendicular to each other.

____

5. What symbol in the figure tells you that line segments are perpendicular to each other?

____

Name ______________________ Date __________ Class __________

LESSON **7-2**

Puzzles, Twisters & Teasers

Line Up!

Decide whether the lettered lines in each figure are parallel or perpendicular. Each answer has a corresponding letter. Circle the letter above your answer. Use the letters to solve the riddle.

1.

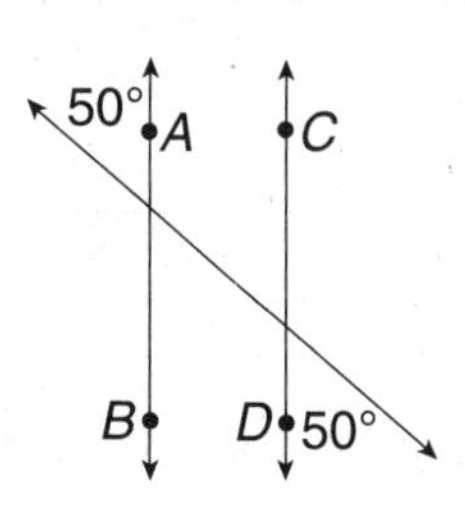

R	**J**
parallel	perpendicular

2.

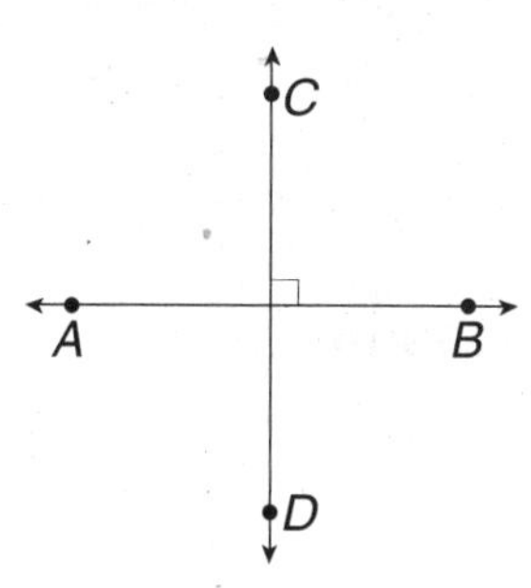

U	**E**
parallel	perpendicular

3.

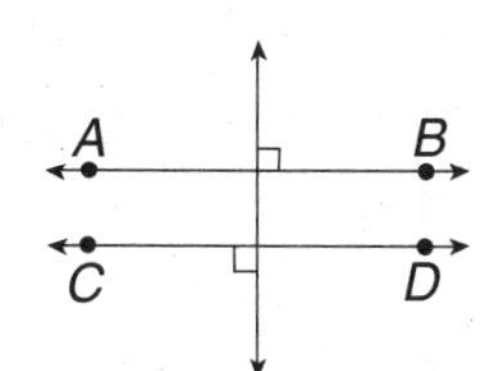

C	**T**
parallel	perpendicular

4.

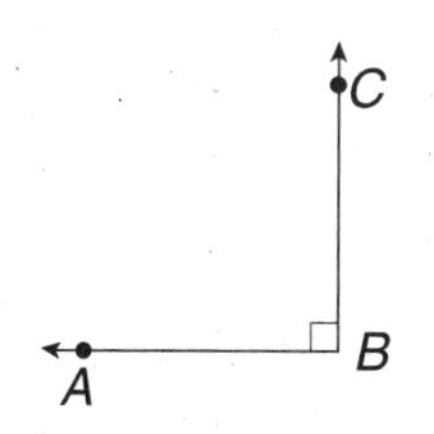

P	**C**
parallel	perpendicular

5.

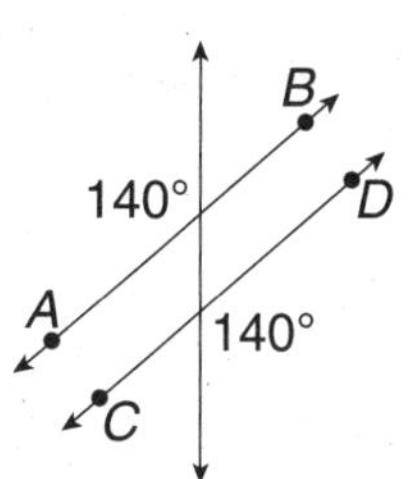

I	**R**
parallel	perpendicular

6.

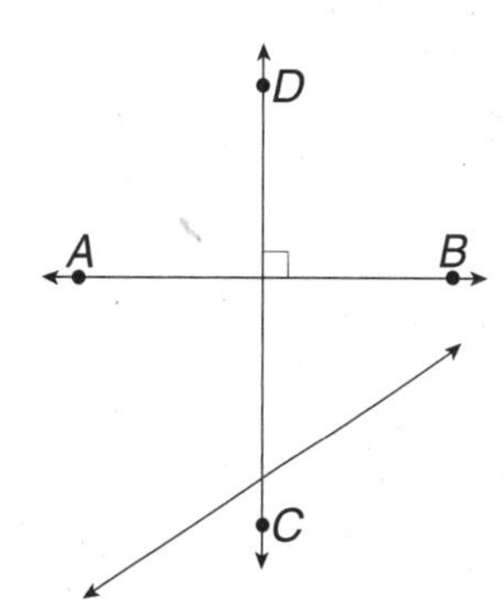

O	**N**
parallel	perpendicular

What do you call it when you ride your bike around and around the block?

____ ____ ____ Y ____ L ____ ____ G

1 2 3 4 5 6

Name ______________________ Date __________ Class __________

LESSON 7-3

Practice A

Angles in Triangles

Identify each triangle by its angles and sides.

1.

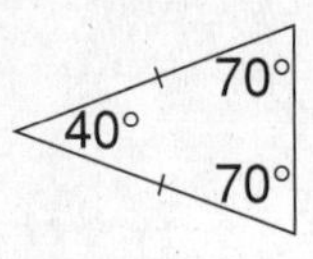

____

2.

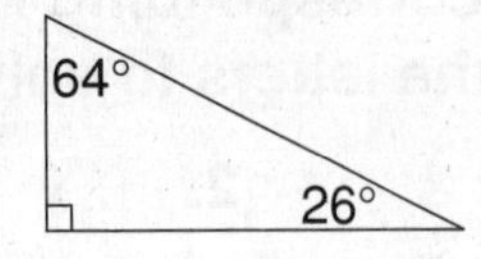

____

3.

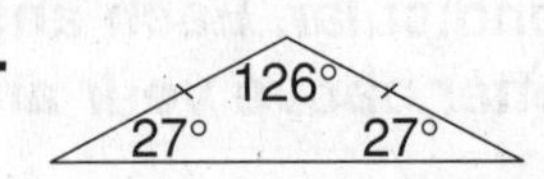

____

Find each angle measure.

4. Find $x°$ in the acute triangle.

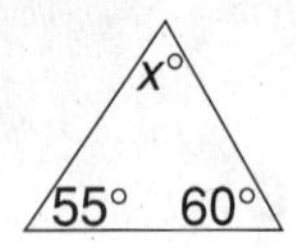

____

5. Find $y°$ in the right triangle.

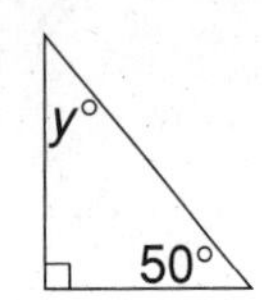

____

6. Find $r°$ in the obtuse triangle.

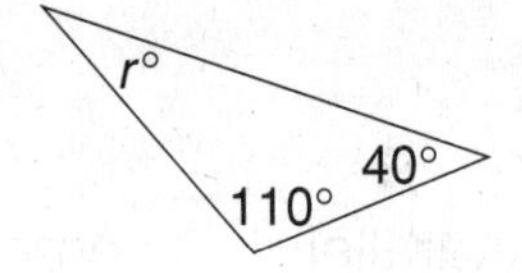

____

7. Find $x°$ in the acute triangle.

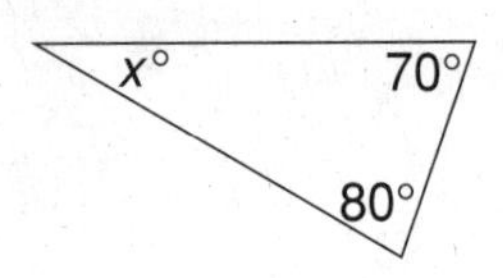

____

8. Find $y°$ in the right triangle.

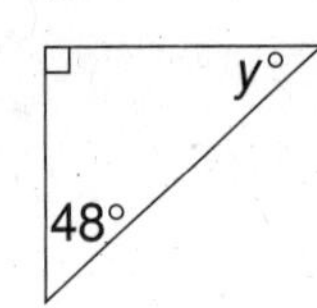

____

9. Find $m°$ in the obtuse triangle.

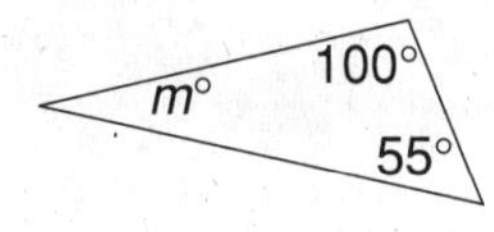

____

10. Find $t°$ in the isoceles triangle.

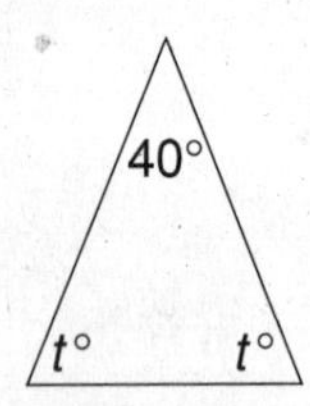

____

11. Find $x°$ in the scalene triangle.

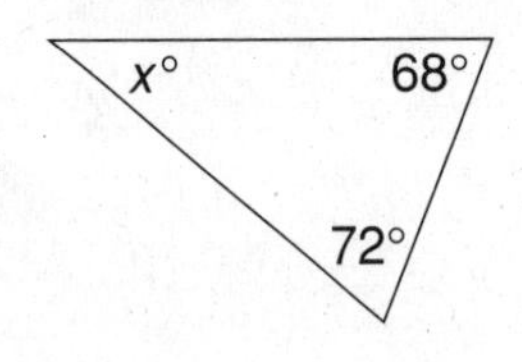

____

12. Find $n°$ in the isoceles triangle.

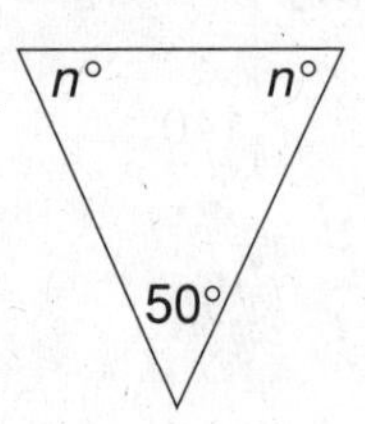

____

13. The second angle in a triangle is three times as large as the first. The third angle is one third as large as the second angle. Find the angle measures. Draw a possible picture of the triangle.

____

Name ______________________________ Date ____________ Class ____________

LESSON 7-3

Practice B

Angles in Triangles

1. Find $x°$ in the right triangle.

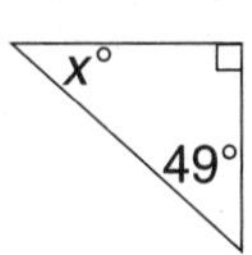

____

2. Find $y°$ in the obtuse triangle.

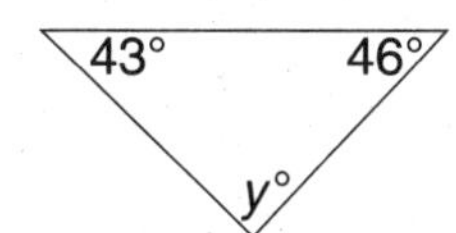

____

3. Find $m°$ in the acute triangle.

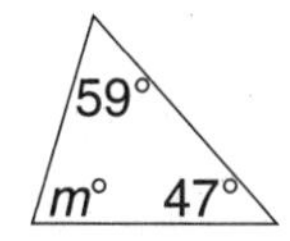

____

4. Find $n°$ in the obtuse triangle.

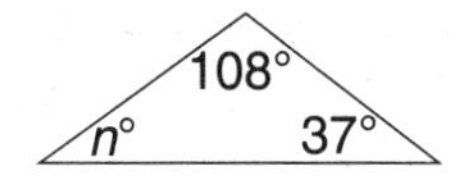

____

5. Find $w°$ in the acute triangle.

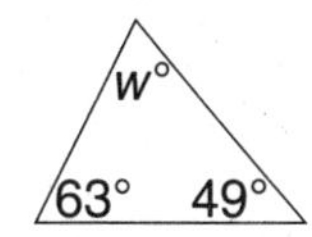

____

6. Find $t°$ in the right triangle.

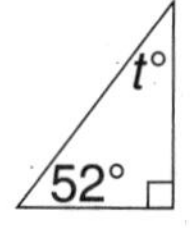

____

7. Find $t°$ in the scalene triangle.

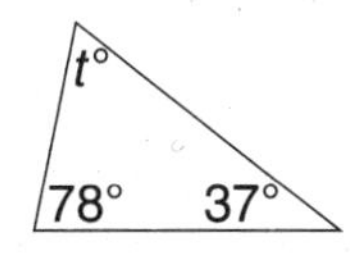

____

8. Find $x°$ in the isosceles triangle.

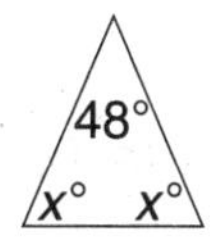

____

9. Find $n°$ in the scalene triangle.

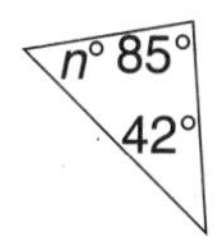

____

10. Find $x°$ in the isosceles triangle.

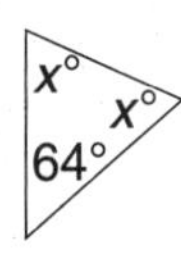

____

11. Find y in the equilateral triangle.

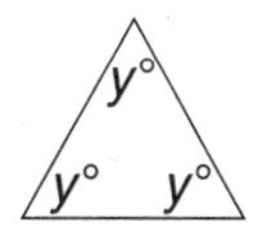

____

12. Find r in the isoceles triangle.

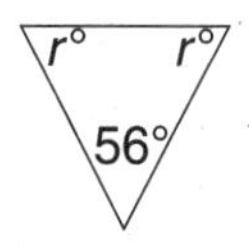

____

13. The second angle in a triangle is one third as large as the first. The third angle is two thirds as large as the first angle. Find the angle measures. Draw a possible picture of the triangle.

____

Name ______________________ Date ____________ Class ____________

LESSON **7-3**

Practice C

Angles in Triangles

Find the value of each variable.

1.

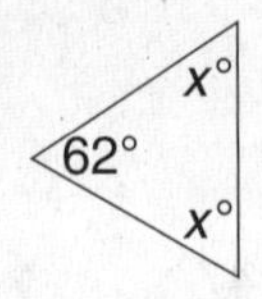

____

2.

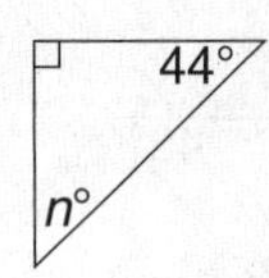

____

3.

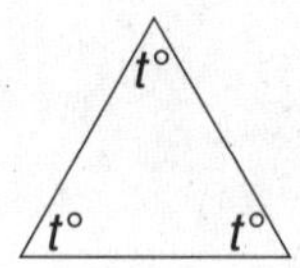

____

4.

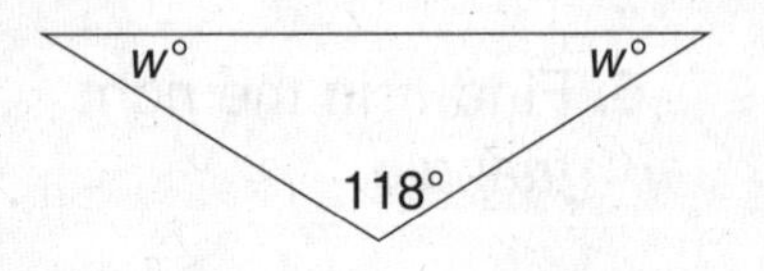

____

5.

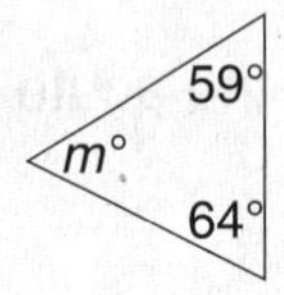

____

6.

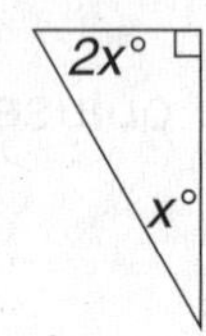

____

7. The measure of the second angle in a triangle is four more than the measure of the first angle and the measure of the third angle is eight more than twice the measure of the first angle. Find the measure of each angle.

____

Describe each statement as *always*, *sometimes*, or *never* true.

8. An obtuse triangle is a scalene triangle.

____

9. A scalene triangle is an isosceles triangle.

____

10. A right triangle is an isosceles triangle.

____

11. A triangle with all angles congruent is acute.

____

Sketch a triangle to fit each description. If no triangle can be drawn, write *not possible*.

12. obtuse, isosceles

13. right, scalene

14. acute, equilateral

Name ______________________ Date ___________ Class ___________

LESSON 7-3

Reteach

Angles in Triangles

Acute Triangle | **Right Triangle** | **Obtuse Triangle**

3 acute angles

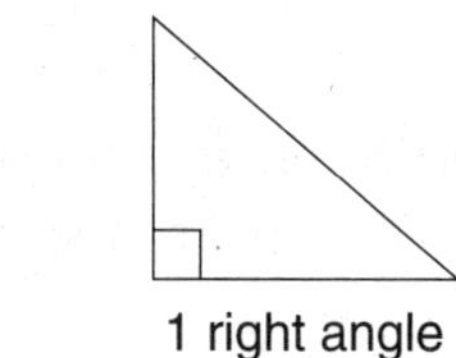
1 right angle

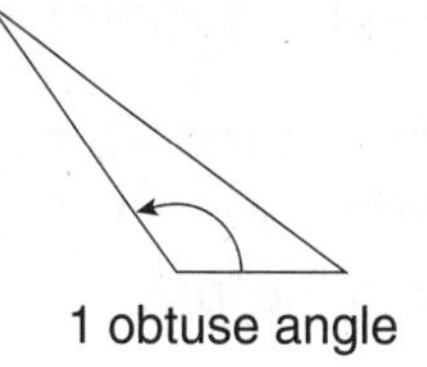
1 obtuse angle

Triangle Sum Theorem: The sum of the measures of the three interior angles of any triangle is 180°.

In the diagram: $m\angle A + m\angle B + m\angle C = 180°$

$a° + b° + c° = 180°$

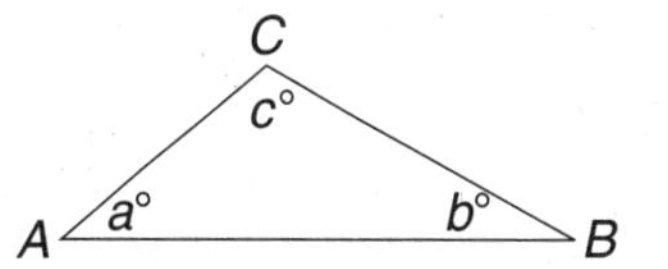

Complete to find the measure of the unknown angle.

1.

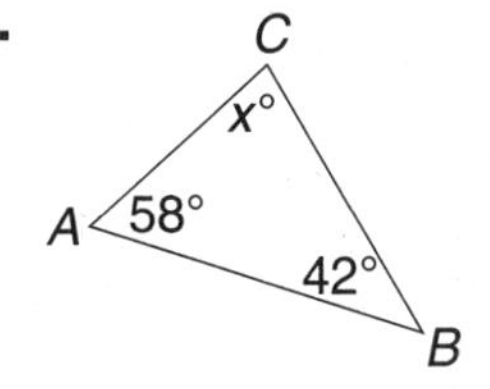

$m\angle A + m\angle B + m\angle C = 180°$

_______ + _______ + $x°$ = _______

_______________ + $x°$ = _______

$x°$ = _______

2.

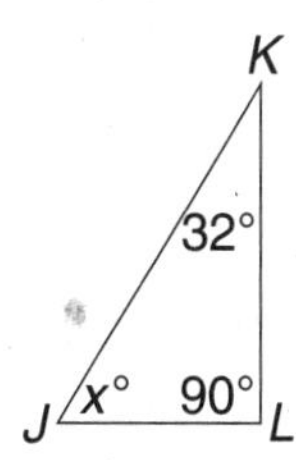

$m\angle J + m\angle K + m\angle L = 180°$

$x°$ + _______ + _______ = _______

$x°$ + _______________ = _______

$x°$ = _______

3.

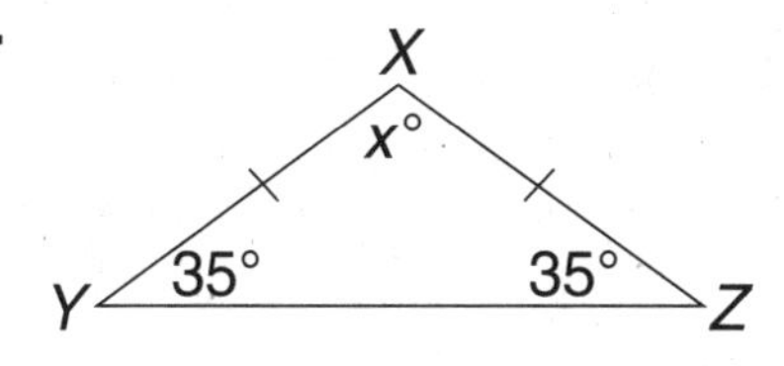

$m\angle X + m\angle Y + m\angle Z = 180°$

____

____

$x°$ = _______

4.

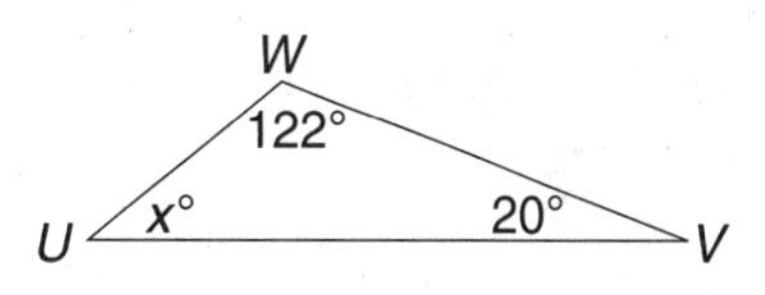

$m\angle U + m\angle V + m\angle W = 180°$

____

____

$x°$ = _______

Name ______________________ Date __________ Class __________

LESSON 7-3

Challenge

Change a This into a That

A **geometric dissection** involves cutting a figure into pieces that can then be rearranged to form another figure.

Trace each figure. Cut up the figure you have traced and rearrange the numbered pieces to form the indicated figure. Sketch your solution.

1. Rearrange the pieces of the equilateral triangle to form a square.

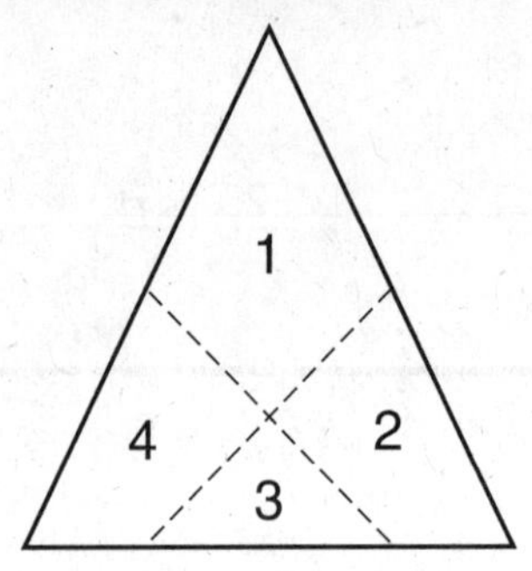

2. Rearrange the pieces of the star to form an equilateral triangle.

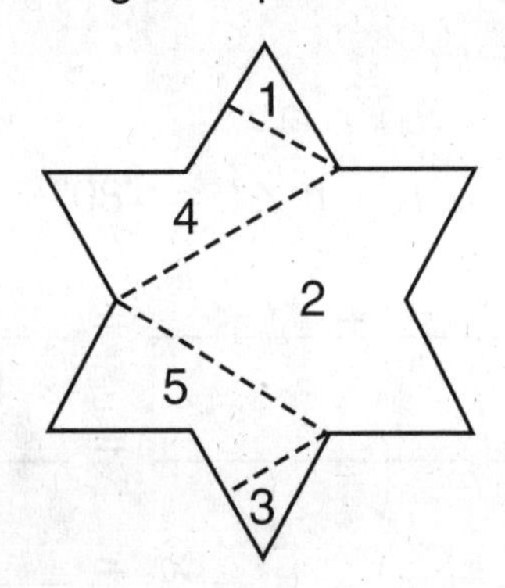

3. Rearrange the pieces of the cross to form an equilateral triangle.

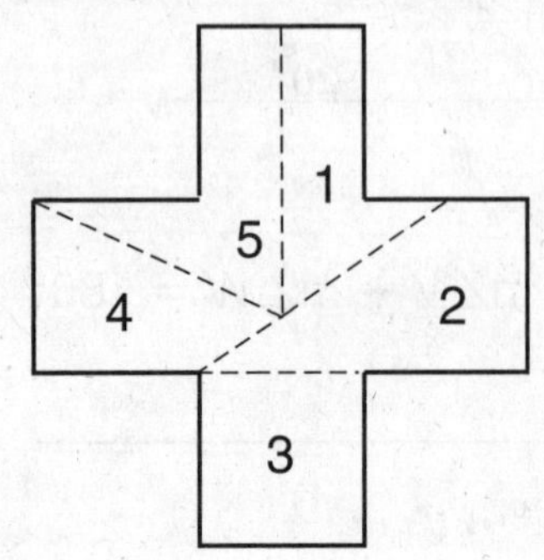

Name ______________________ Date __________ Class __________

LESSON 7-3

Problem Solving

Angles in Triangles

The American flag must be folded according to certain rules that result in the flag being folded into the shape of a triangle. The figure shows a frame designed to hold an American flag.

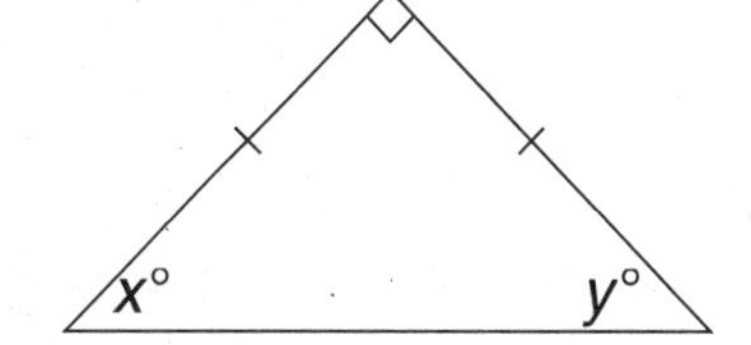

1. Is the triangle acute, right, or obtuse?

____

2. Is the triangle equilateral, isosceles, or scalene?

____

3. Find $x°$.

____

4. Find $y°$.

____

The figure shows a map of three streets. Choose the letter for the best answer.

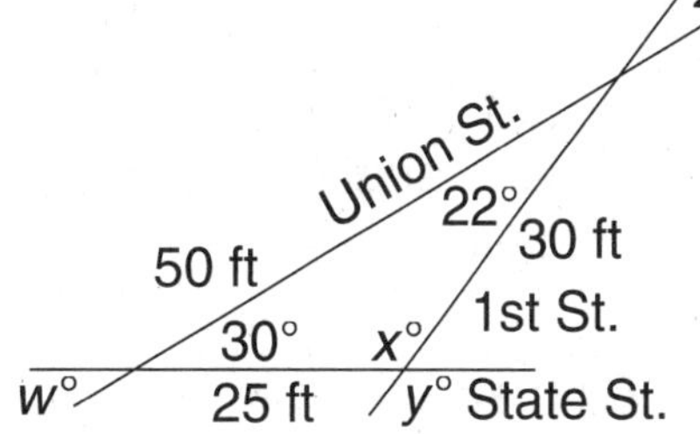

5. Find $x°$.
   - A 22°
   - B 128°
   - C 30°
   - D 68°

6. Find $w°$.
   - F 22°
   - G 128°
   - H 30°
   - J 52°

7. Find $y°$.
   - A 22°
   - B 30°
   - C 128°
   - D 143°

8. Find $z°$.
   - F 22°
   - G 30°
   - H 128°
   - J 143°

9. Which word best describes the triangle formed by the streets?
   - A acute
   - B right
   - C obtuse
   - D equilateral

10. Which word best describes the triangle formed by the streets?
    - F equilateral
    - G isosceles
    - H scalene
    - J acute

Name ______________________________ Date ____________ Class ____________

LESSON 7-3

Reading Strategies

Graphic Organizer

A triangle can be classified by the measurement of one or more of its angles.

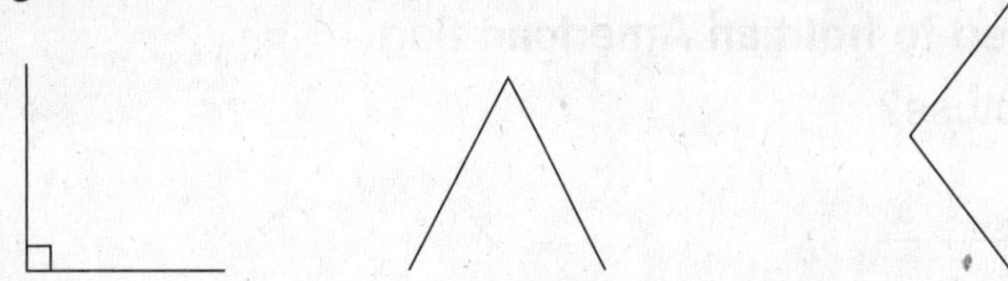

Right angle $= 90^\circ$ Acute angle $< 90^\circ$ Obtuse angle $> 90^\circ$

A triangle can also be classified by the length of its sides. This chart will help you compare triangles by angles and sides.

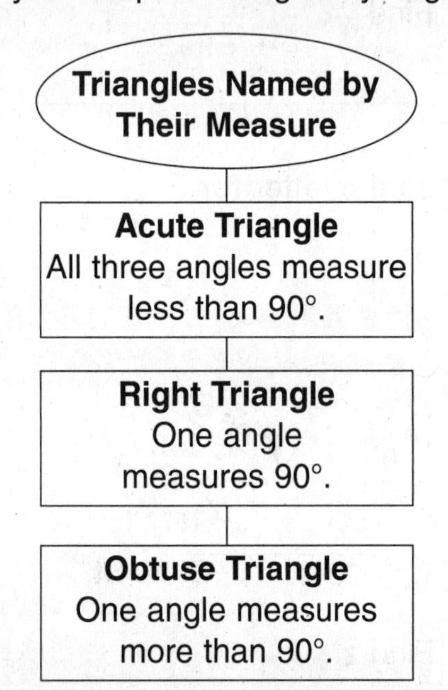

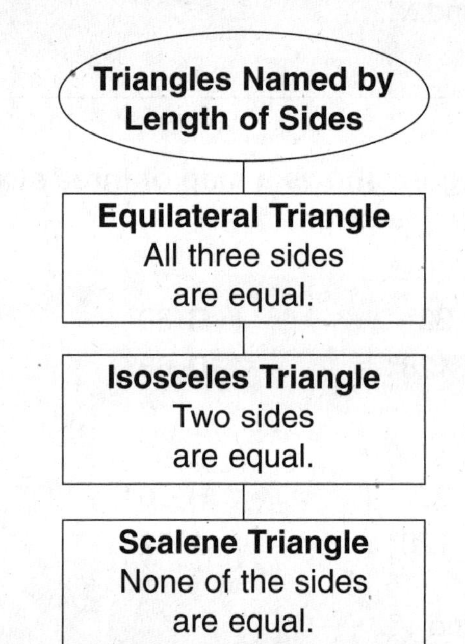

Write two of the following terms to describe each triangle: acute, right, obtuse, equilateral, isosceles, scalene.

1.

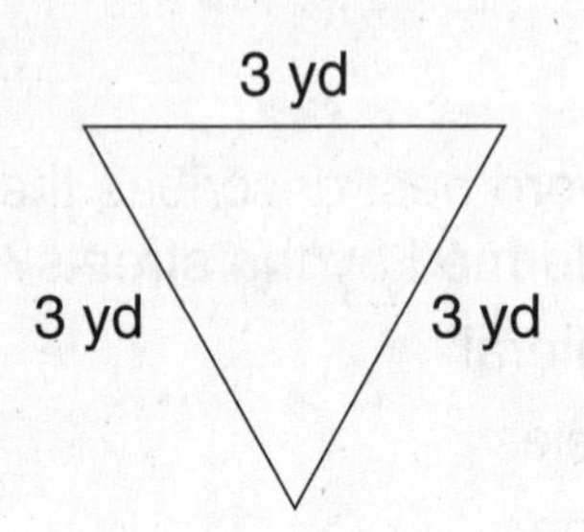

_____ ________________

2.

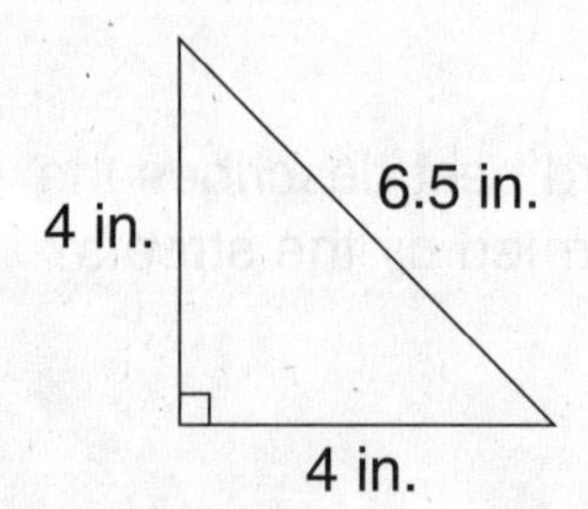

____

Name ______________________________ Date ____________ Class ____________

LESSON 7-3

Puzzles, Twisters & Teasers

Try Some Triangles

Across

1. A ______ triangle has one 90° angle.

5. An ______ triangle has 3 acute angles.

6. The Triangle Sum ______ says that the 3 angles in a triangle always equal 180°.

7. A ______ triangle has no congruent sides and no congruent angles.

Down

2. An ______ triangle has at least 2 congruent sides and 2 congruent angles.

3. An ______ triangle has 3 congruent sides and 3 congruent angles.

4. An ______ triangle has 1 obtuse angle.

Name ______________________________ Date ____________ Class ____________

LESSON 7-4

Practice A

Classifying Polygons

Name each polygon.

1. ____________

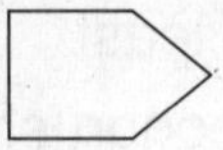

2. ____________

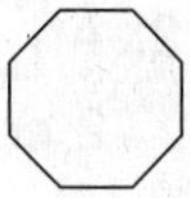

3. ____________

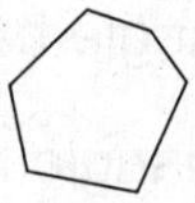

Find the sum of the angle measures in each figure.

4. ____________

5. ____________

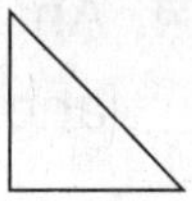

6. ____________

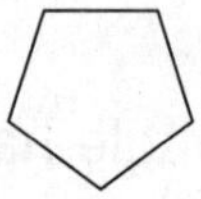

7. ____________

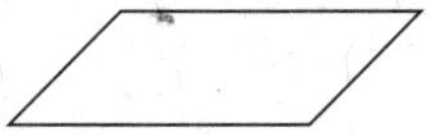

8. ____________

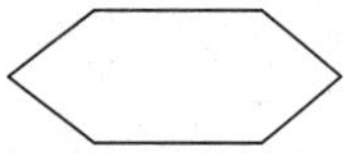

9. ____________

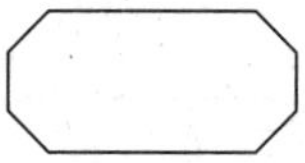

Find the angle measures in each regular polygon.

10. ____________

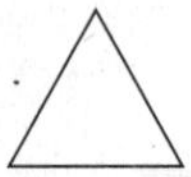

11. ____________

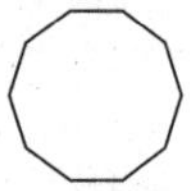

12. ____________

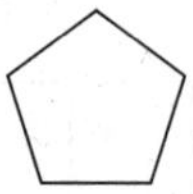

13. ____________

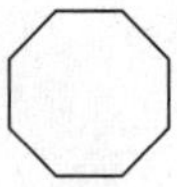

14. ____________

15. ____________

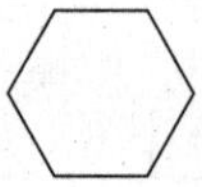

Write all the names that apply to each figure.

16. ____________

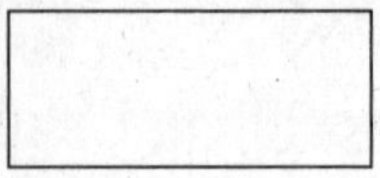

17. ____________

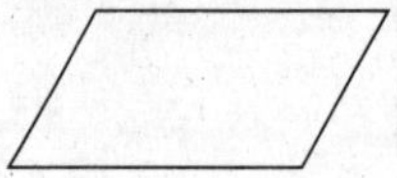

18. ____________

Name ______________________ Date __________ Class __________

LESSON 7-4

Practice B

Classifying Polygons

Find the sum of the angle measures in each figure.

1. 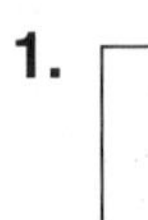

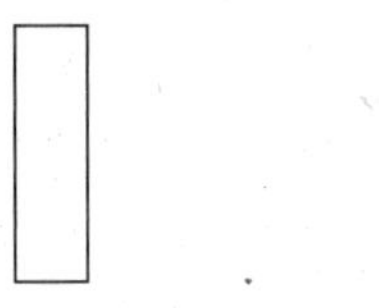

____

2. 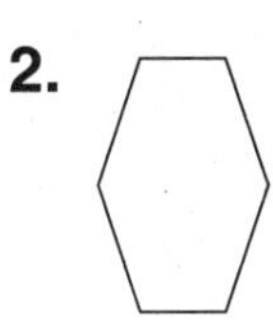

____

3. 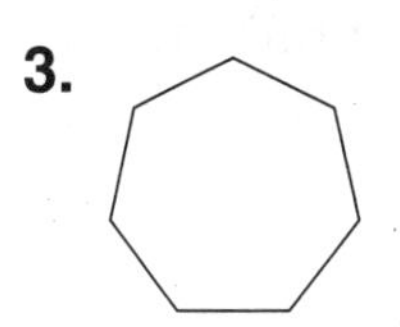

____

4. 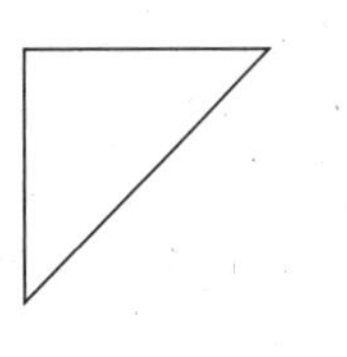

____

5.

____

6. 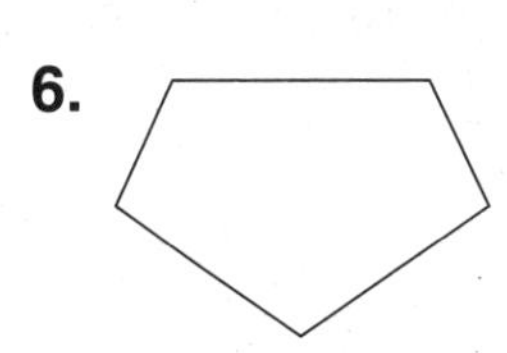

____

Find the angle measures in each regular polygon.

7. 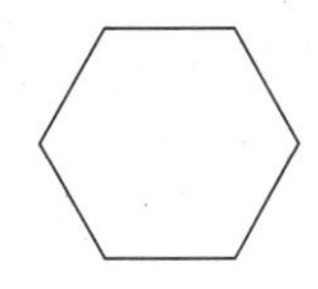

____

8. 

____

9. 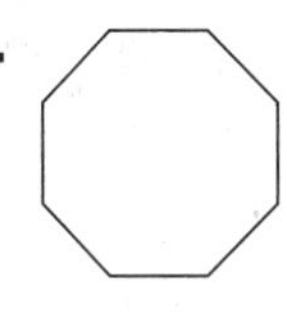

____

10. 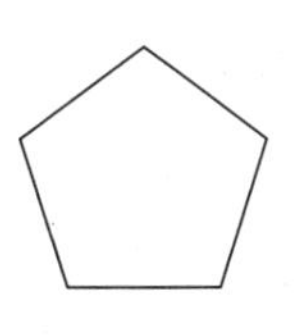

____

11. 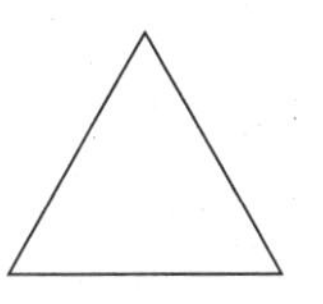

____

12. 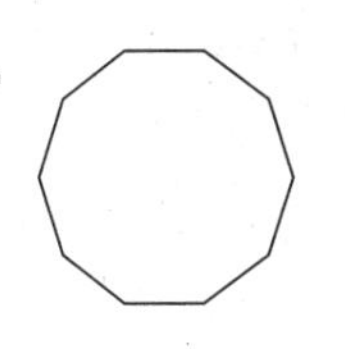

____

Give all the names that apply to each figure.

13. 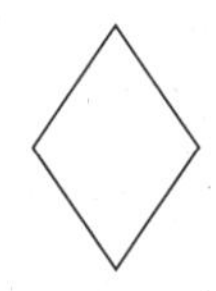

____

____

14. 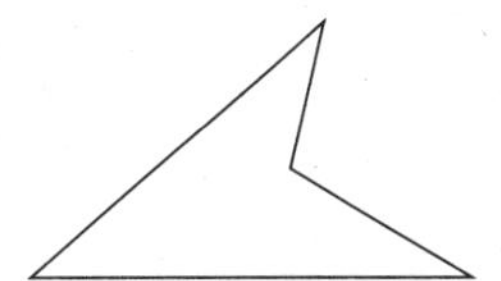

____

____

15. 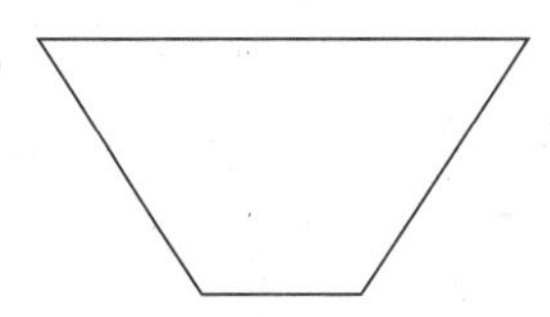

____

____

Name ______________________ Date __________ Class __________

LESSON 7-4

Practice C
Classifying Polygons

Find the sum of the angle measures in each regular polygon. Then, find the measure of each angle.

1. 24-gon

2. 16-gon

3. 36-gon

______________ ______________ ______________

Find the value of each variable.

4.

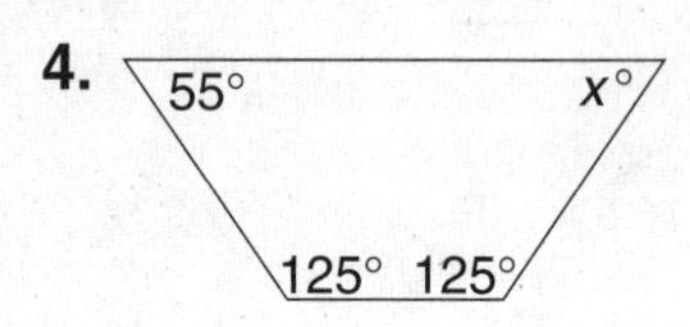

5.

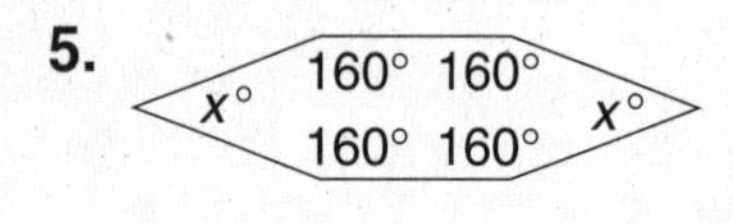

6.

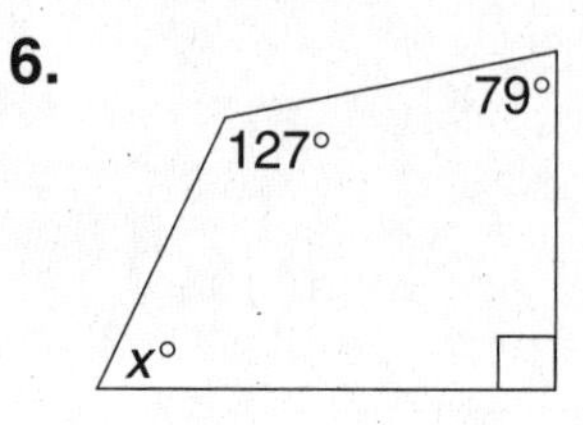

______________ ______________ ______________

The sum of the angle measures of a polygon is given. Name the polygon.

7. 1080°

8. 1260°

9. 900°

10. 1440°

______________ ______________ ______________ ______________

Graph the given vertices on a coordinate plane. Connect the points to draw a polygon and classify it by the number of its sides.

11. (1, 5), (4, 2), (4, −2), (1, −5), (−3, −5), (−5, −2), (−5, 2), (−3, 5)

____

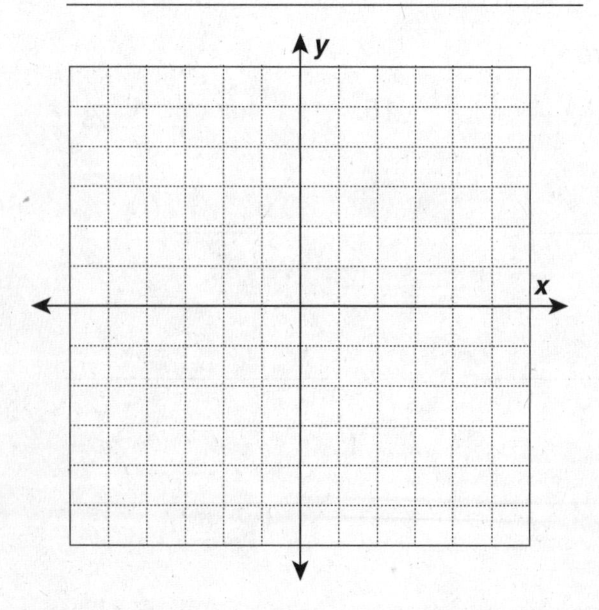

12. (0, −1), (−1, 3), (2, 5), (5, 3), (4, −1)

____

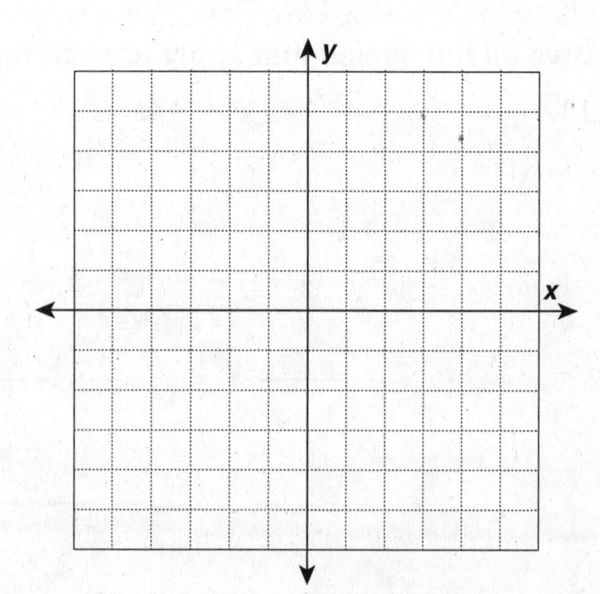

Name ______________________ Date __________ Class __________

LESSON 7-4

Reteach

Classifying Polygons

A polygon of n sides (an ***n*-gon**) can be divided into $(n - 2)$ triangles
The sum of the angle measures of an n-gon $= (n - 2)180°$.

A polygon of 5 sides (pentagon) can be divided into 3 triangles.

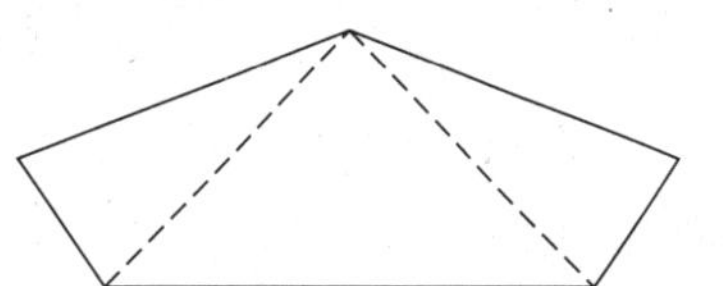

Sum of angle measures of pentagon
$= (n - 2)\ 180°$
$= (5 - 2)\ 180° = (3)180° = 540°$

Find the sum of the measures of the angles.

	1. quadrilateral	2. hexagon
How many sides in the polygon?	___	___
How many triangles can be formed?	___ − 2 = ___	___ − 2 = ___
Multiply the number of triangles by 180°.	180° × ___	180° × ___
sum of the measures of the angles	______	______

In a **regular polygon,** all sides and all angles are congruent.

The measure of each angle of a regular polygon $= \frac{\text{sum of the angles}}{\text{number of sides}}$

The measure of each angle of a regular pentagon $= \frac{(5 - 2)180°}{5} = \frac{(3)180°}{5} = \frac{540°}{5} = 108°$

Find the measure of each angle.

	3. regular octagon	4. regular decagon
How many sides (angles) in the polygon?	___	___
How many triangles can be formed?	___ − 2 = ___	___ − 2 = ___
Multiply the number of triangles by 180°.	180° × ___	180° × ___
Sum of the measures of the angles	______	______
Divide the sum by the number of angles.	______	______
Measure of each angle of the polygon	______	______

Name ______________________ Date ___________ Class ___________

LESSON 7-4

Reteach

Classifying Polygons (continued)

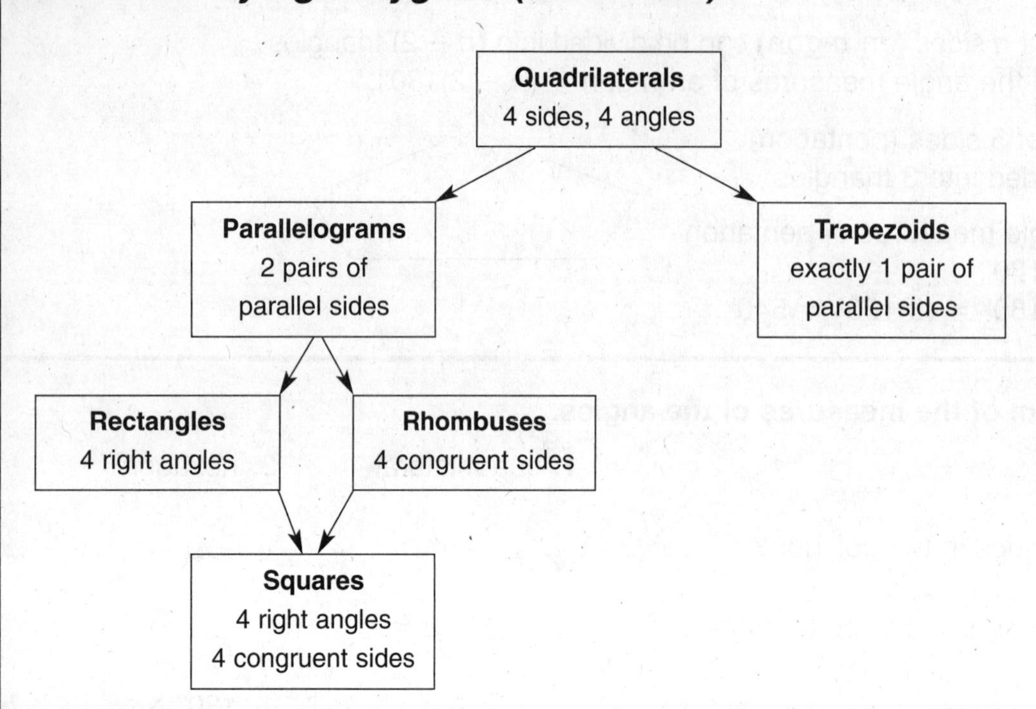

A figure on a lower branch of the tree has the properties of the figures above it.

All rectangles are parallelograms.
But not all parallelograms are rectangles.

Write all the names that apply to each figure.

5.

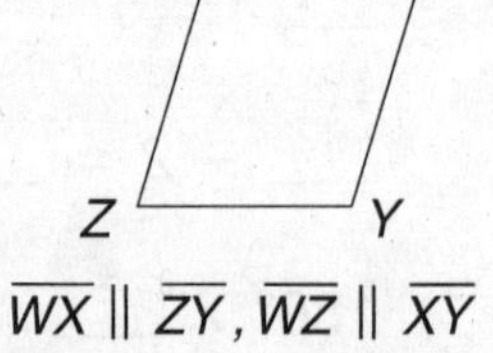

$\overline{AB} \parallel \overline{DC}$

four-sided polygon ____________

1 pair of parallel sides ____________

6.

$\overline{WX} \parallel \overline{ZY}, \overline{WZ} \parallel \overline{XY}$

$\overline{WX} \cong \overline{XY} \cong \overline{ZY} \cong \overline{WZ}$

four-sided polygon ____________

2 pairs of parallel sides ____________

4 congruent sides ____________

7.

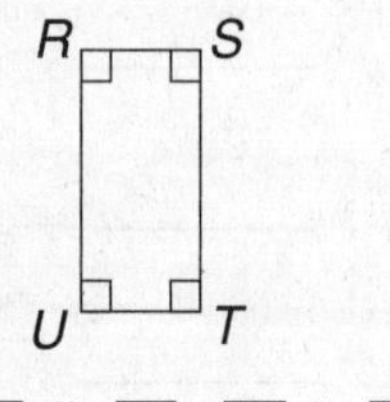

$\overline{RS} \parallel \overline{UT}, \overline{RU} \parallel \overline{ST}$

four-sided polygon ____________

2 pairs of parallel sides ____________

4 right angles ____________

Name ______________________ Date __________ Class __________

LESSON 7-4

Challenge
Slanted View

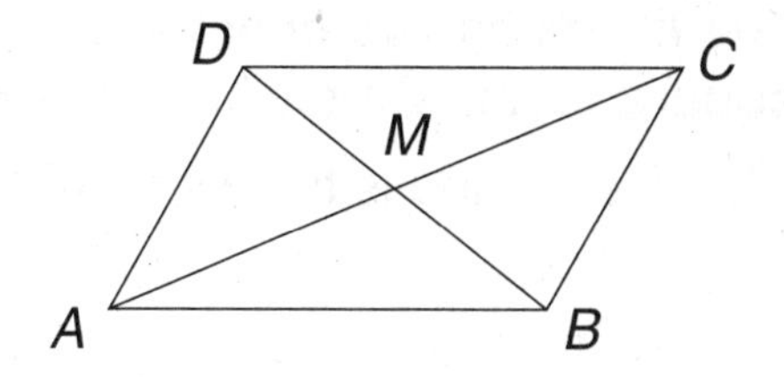

1. Refer to parallelogram *ABCD*. Use a ruler.
   a. Is diagonal $\overline{AC} \cong$ diagonal $\overline{BD}$? ____________
   b. Is $\overline{AM} \cong \overline{MC}$? Is $\overline{DM} \cong \overline{MB}$? ____________

   Make a statement about how the diagonals of a parallelogram relate to each other.

   ____

   ____

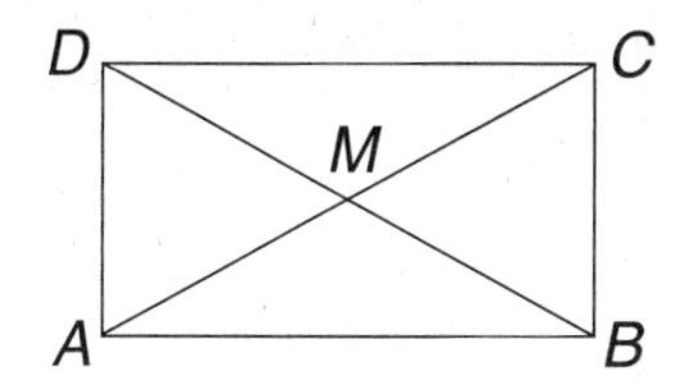

2. Refer to rectangle *ABCD* and your observations in Question 1.
   a. Since a rectangle is a parallelogram, what property should the diagonals have? Use a ruler to verify your conjecture.

   ____

   b. What additional property do the diagonals of a rectangle have?

   ____

3. Refer to rhombus *ABCD*.

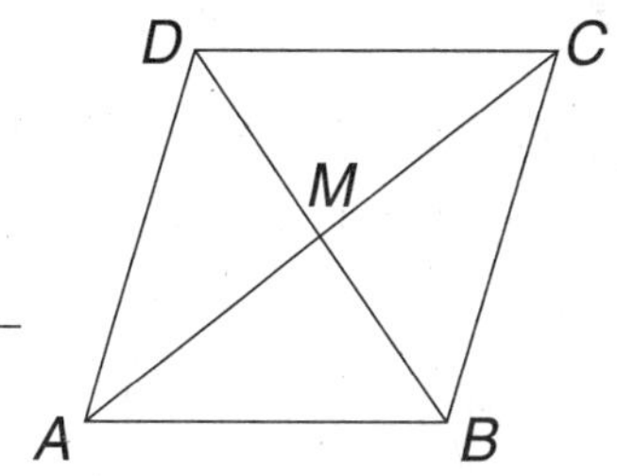

   a. Since a rhombus is a parallelogram, what property should the diagonals have? Use a ruler to verify your conjecture.

   ____

   b. Are the diagonals congruent? ____________
   c. Measure the angles with vertex *M*. What additional property do the diagonals of a rhombus have?

   ____

   d. Measure the angles at each vertex of the rhombus. What additional property do the diagonals of a rhombus have?

   ____

4. Make a conjecture about the properties of the diagonals of a square. Draw a square and verify your conjectures with a ruler and protractor.

   ____

   ____

Name ______________________ Date ____________ Class ____________

LESSON **7-4**

Problem Solving

Classifying Polygons

The figure shows how the glass for a window will be cut from a square piece. Cuts will be made along $\overline{CE}$, $\overline{FH}$, $\overline{IK}$, and $\overline{LB}$.

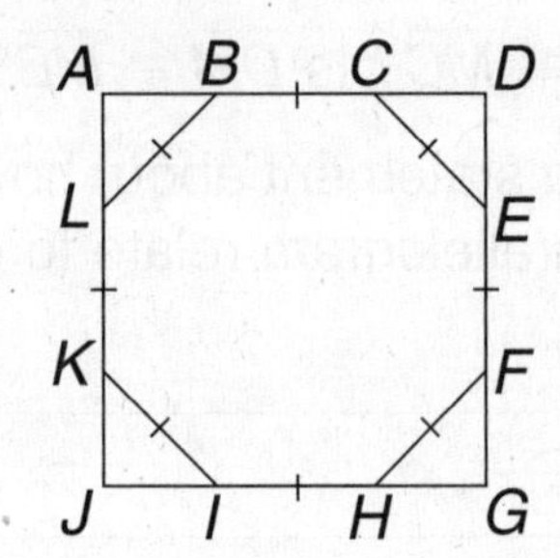

1. What shape is the window?

____

2. What is the sum of the angle measures of the window?

____

3. What is the measure of each angle of the window?

____

4. Based on the angles, what kind of triangle is $\triangle CDE$?

____

5. Based on the sides, what kind of triangle is $\triangle CDE$?

____

The figure shows how parallel cuts will be made along $\overline{AD}$ and $\overline{BC}$. $\overline{AB}$ and $\overline{CD}$ are parallel. Choose the letter for the best answer.

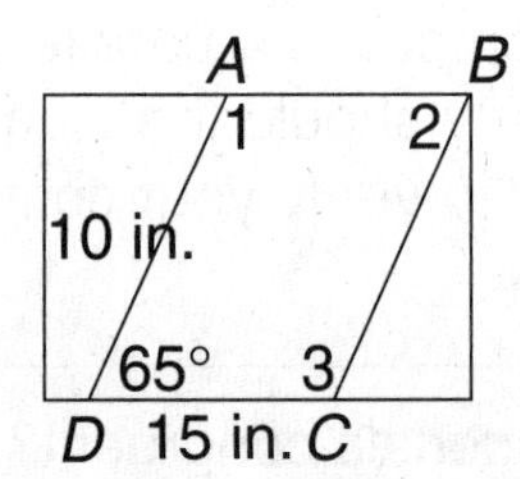

6. Which word correctly describes figure *ABCD* after the cuts are made?

A triangle

B quadrilateral

C pentagon

D hexagon

7. Which word correctly describes figure *ABCD* after the cuts are made?

F parallelogram

G trapezoid

H rectangle

J rhombus

8. Find the measure of ∠1.

A 45°

B 65°

C 90°

D 115°

9. Find the measure of ∠2.

F 45° **H** 65°

G 90° **J** 115°

10. Find the measure of ∠3.

A 45° **C** 65°

B 90° **D** 115°

Name ______________________ Date __________ Class __________

LESSON 7-4

Reading Strategies

Reading a Chart

The prefix *poly-* means "many." The word *polygon* means "many-sided figure." A **polygon** is a closed, plane figure named by the number of its sides.

Polygons **Not Polygons**

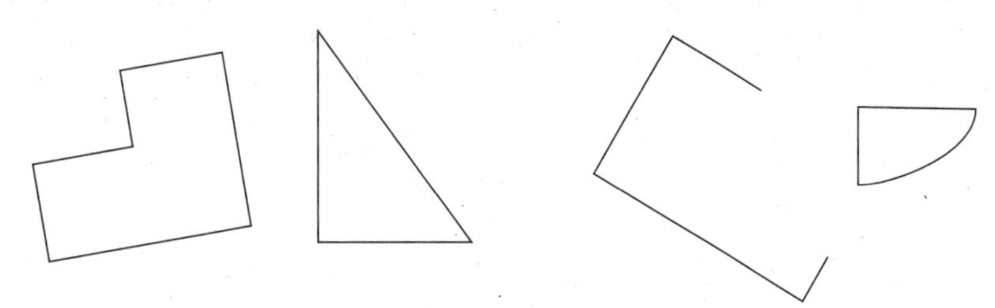

The prefix of a polygon's name tells you the number of its sides and angles.

- *Tri-* means "three" → A **tri**cycle has 3 wheels.
  → A **tri**angle has 3 sides and 3 angles.
- *Quad-* means "four" → A **quad**ruped is a 4-legged animal.
  → A **quad**rilateral has 4 sides and 4 angles.

This chart helps you organize polygons by their sides and angles.

Polygon	Number of Sides	Number of Angles
Triangle	3	3
Quadrilateral	4	4
Pentagon	5	5
Hexagon	6	6
Octagon	8	8

Write the name of each polygon. Use the chart to help you.

1.

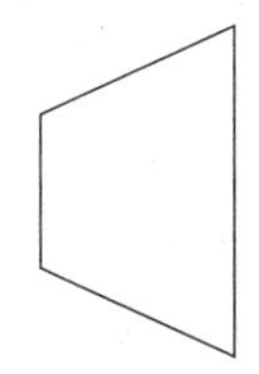

____

2.

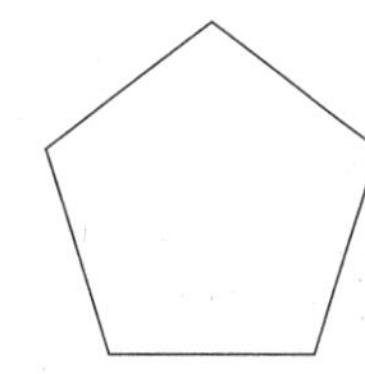

____

3.

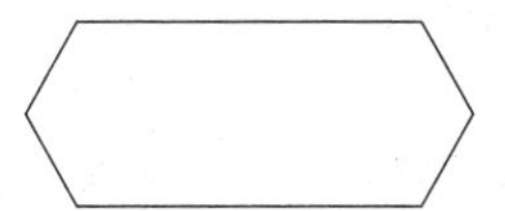

____

4.

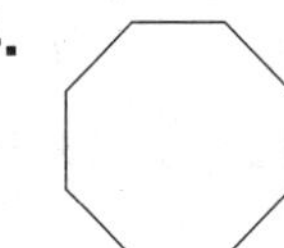

____

Name ________________________ Date ____________ Class ____________

LESSON 7-4

Puzzles, Twisters & Teasers

What Side Are You On?!

Name each figure, one letter on each space. Each answer has one or two boxed letters. Unscramble the boxed letters to solve the riddle.

1.

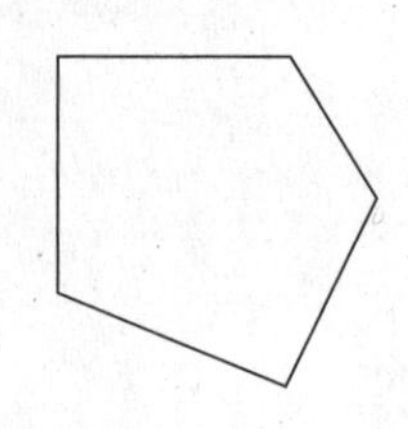

___ ___ ___ ___ [ ] ___ ___ ___

2.

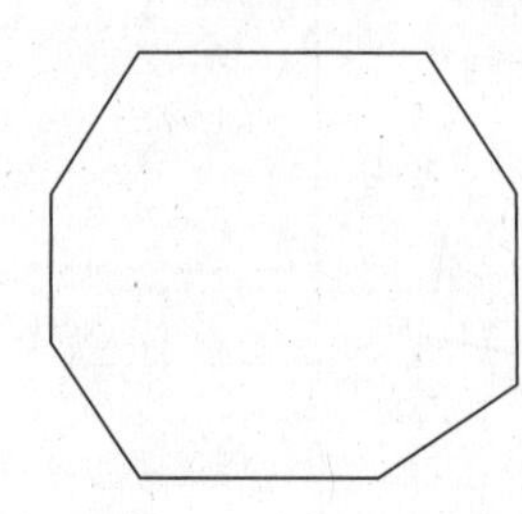

___ ___ ___ ___ [ ] ___ ___ ___

3.

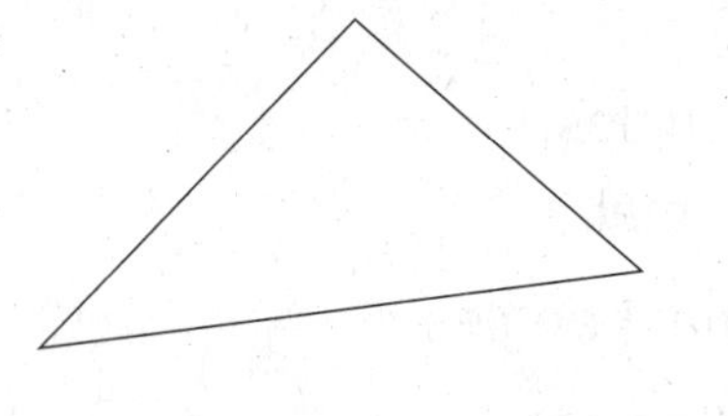

___ ___ [ ] ___ ___ ___ ___ ___

4.

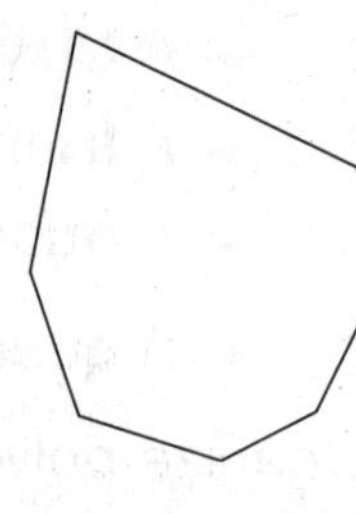

[ ] ___ ___ ___ ___ ___ ___ ___

5.

___ ___ ___ ___ [ ] ___ ___ ___ ___ ___ ___ ___ ___ [ ] ___ ___

6.

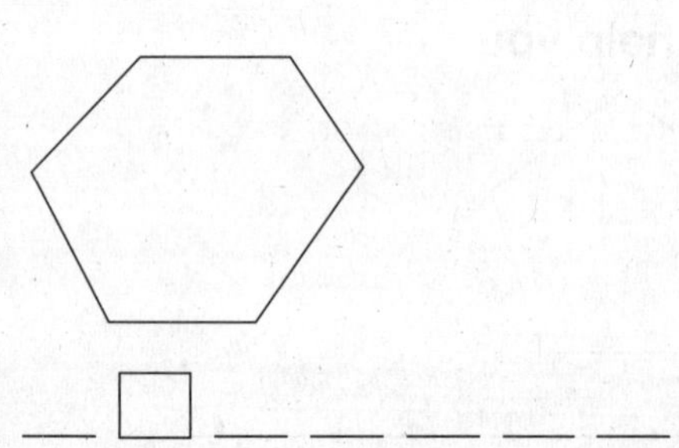

___ [ ] ___ ___ ___ ___ ___

Why was the computer tired after its long trip?

It was ___ ___ ___ R ___ D ___ ___ V ___.

Name ______________________ Date __________ Class __________

LESSON 7-5

Practice A

Coordinate Geometry

Fill in each blank with the correct word from the box at the right.

downward
horizontal
upward
vertical

1. If a line has a positive slope, it slants __________ to the right.

2. If a line has a negative slope, it slants __________ to the right.

3. The slope of a __________ line is undefined.

4. A __________ line has a slope of 0.

Determine if the slope of each line is positive, negative, 0, or undefined. Then find the slope of each line.

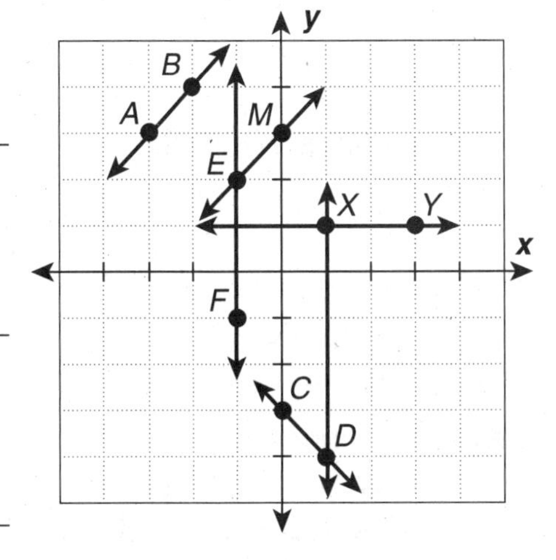

5. $\overleftrightarrow{AB}$ __________

6. $\overleftrightarrow{EF}$ __________

7. $\overleftrightarrow{CD}$ __________

8. $\overleftrightarrow{XY}$ __________

9. $\overleftrightarrow{EM}$ __________

10. $\overleftrightarrow{DX}$ __________

11. Which lines are parallel?

____

12. Which lines are perpendicular?

____

Graph the quadrilateral with the given vertices. Write all the names that apply to the quadrilateral.

13. (–2, 2), (3, 2), (1, –2), (–4, –2)

____

____

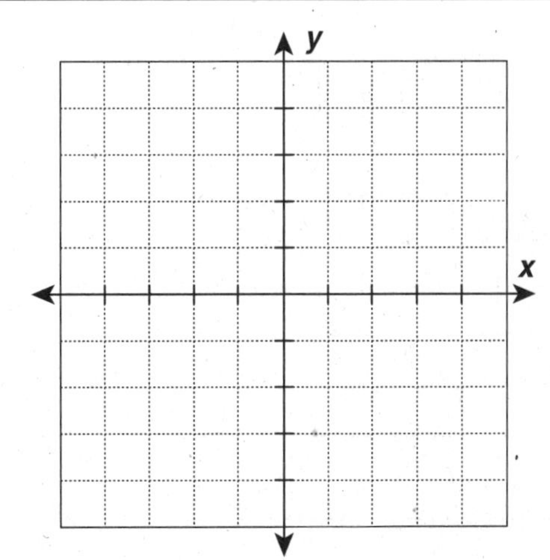

Name ______________________ Date ____________ Class ____________

LESSON 7-5

Practice B

Coordinate Geometry

Determine if the slope of each line is positive, negative, 0, or undefined. Then find the slope of each line.

1. $\overleftrightarrow{AB}$ ______ ______________

2. $\overleftrightarrow{CD}$ ______ ______________

3. $\overleftrightarrow{RS}$ ______ ______________

4. $\overleftrightarrow{TC}$ ______________

5. $\overleftrightarrow{DR}$ ______________

6. $\overleftrightarrow{TX}$ ______________

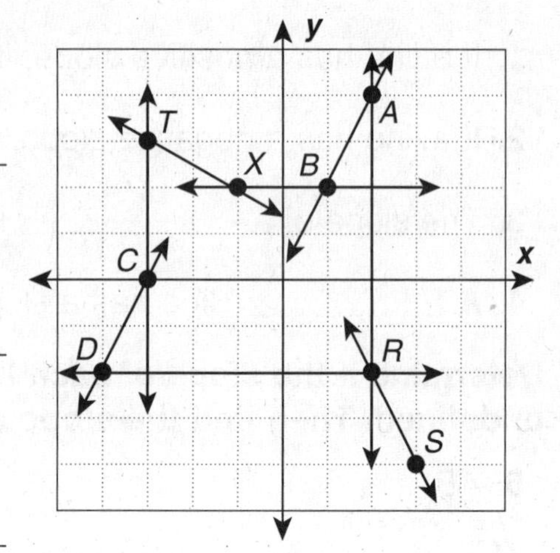

7. Which lines are parallel?

____

8. Which lines are perpendicular?

____

Graph the quadrilateral with the given vertices. Write all the names that apply to the quadrilateral.

9. (−1, 1), (4, 1), (1, −3), (−4, −3)

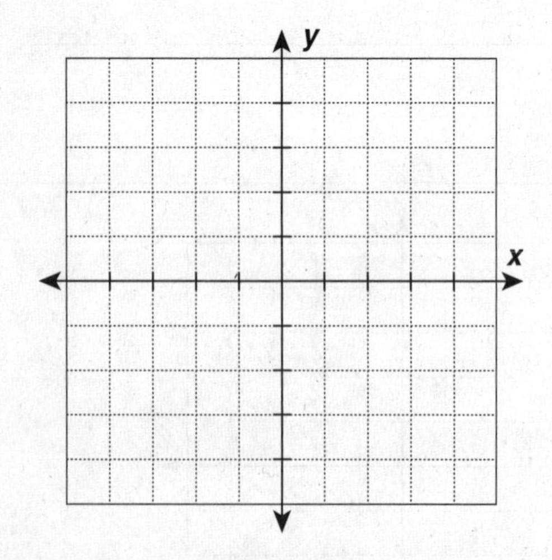

____

____

Find the coordinates of the missing vertex.

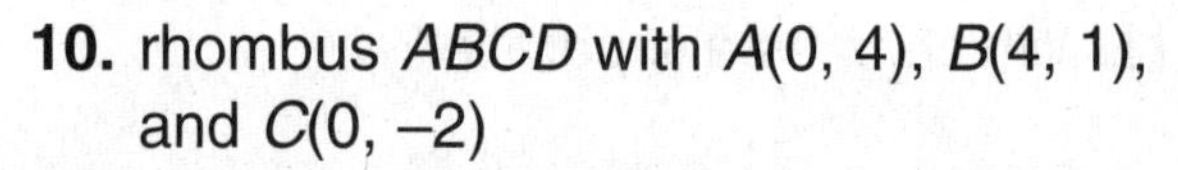

10. rhombus *ABCD* with *A*(0, 4), *B*(4, 1), and *C*(0, –2)

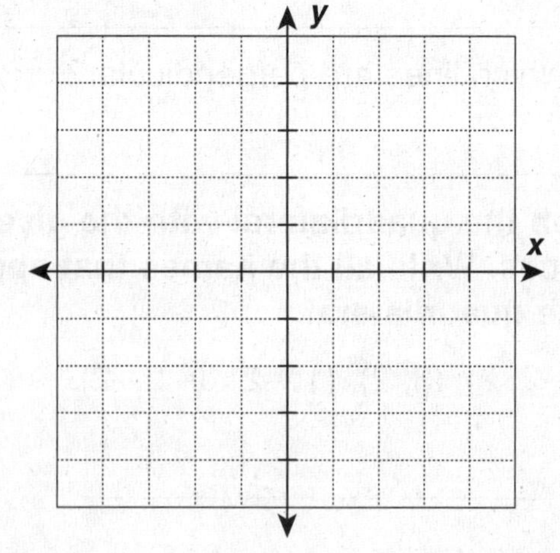

____

Name ______________________ Date __________ Class __________

LESSON 7-5

Practice C

Coordinate Geometry

Draw the line through the given points and find its slope.

1. $A(3, 2)$, $B(4, 4)$

____

2. $C(-2, 1)$, $D(-2, 3)$

____

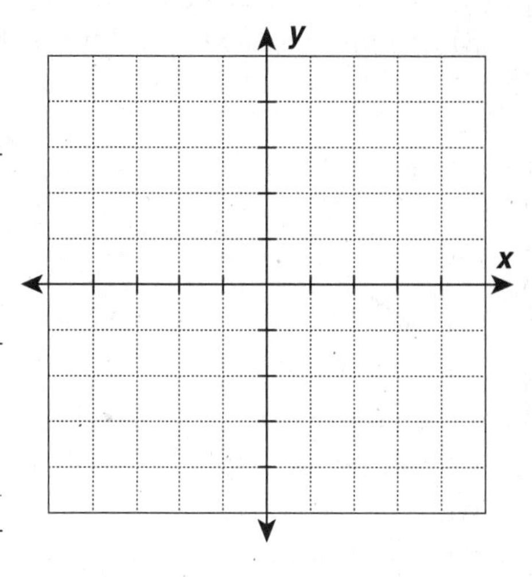

3. $R(-1, -4)$, $S(-3, 4)$

____

4. $X(2, -3)$, $Y(4, -4)$

____

5. $M(-3, -3)$, $N(0, -2)$

____

6. $E(2, 3)$, $F(0, 1)$

____

Find the coordinates of the missing vertex.

7. parallelogram *RSTU* with $R(-4, 4)$, $S(2, 4)$, $T(4, 0)$

____

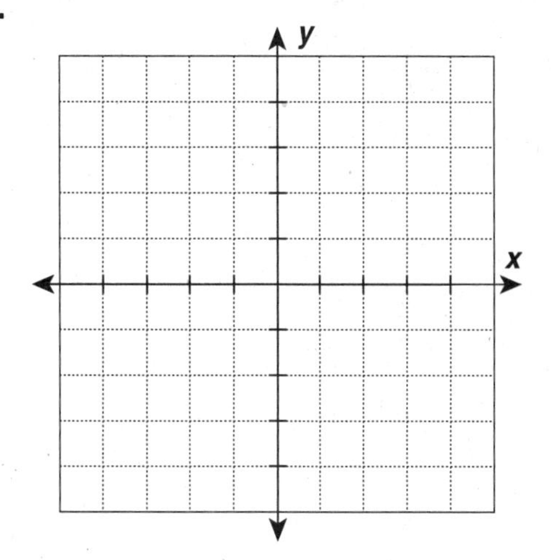

8. On a coordinate grid draw a line *r* with a slope 0 and a line *s* with slope 1. Then draw a line through the intersection of lines *r* and *s* that has a slope between 0 and 1. Name the line and state its slope.

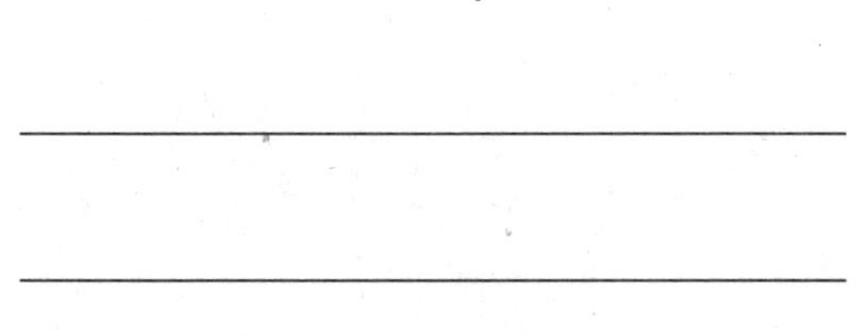

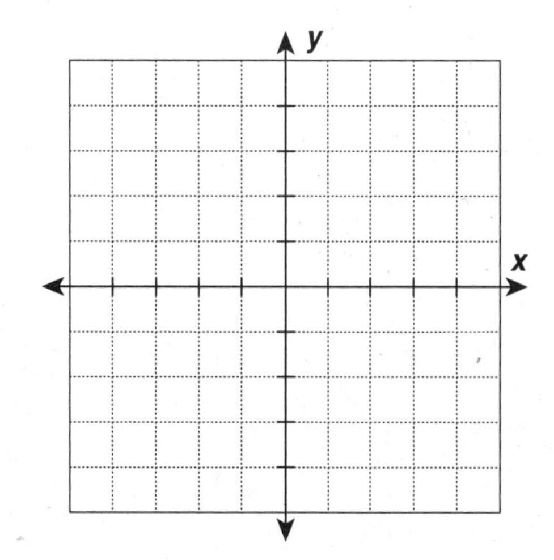

Name ______________________ Date ____________ Class ____________

LESSON 7-5

Reteach
Coordinate Geometry

Possible Values for Slope

Slope is Positive	Slope is Negative	Slope = 0	Slope is Undefined
Line slants up. Forms acute angle with the positive direction of x-axis.	*Line slants down.* Forms obtuse angle with positive direction of x-axis.	*Horizonal Line* Parallel to x-axis.	*Vertical Line* Perpendicular to x-axis

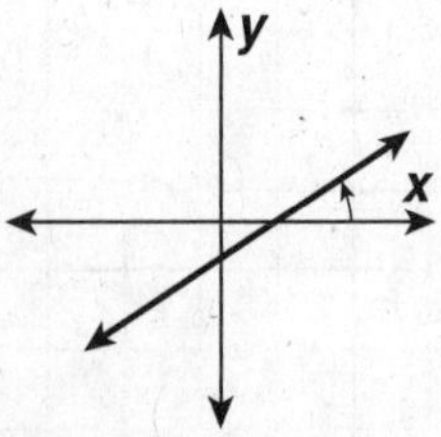

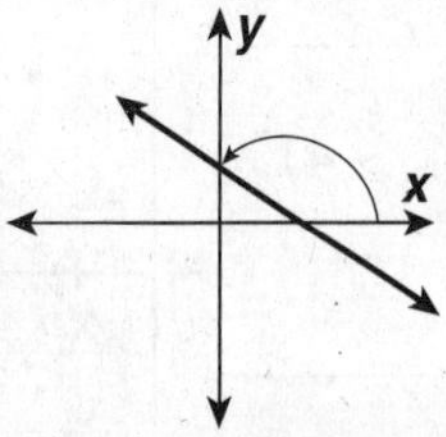

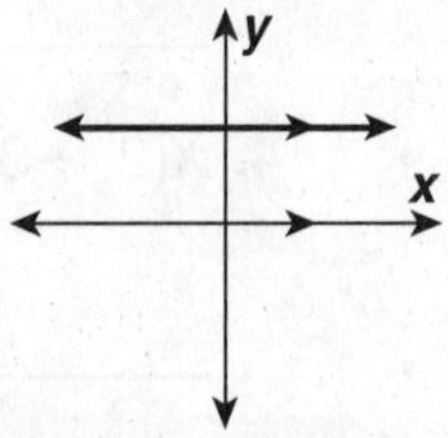

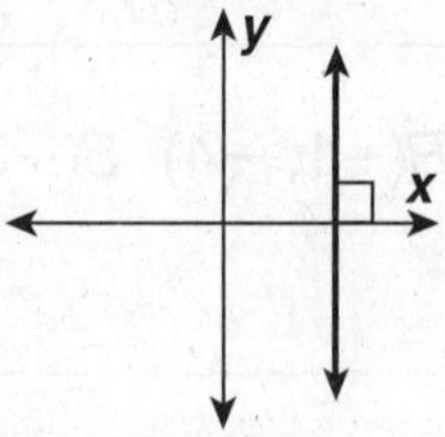

Plot the given points. Describe the slope of the line that joins them.

1. (−2, 2) and (2, 5)

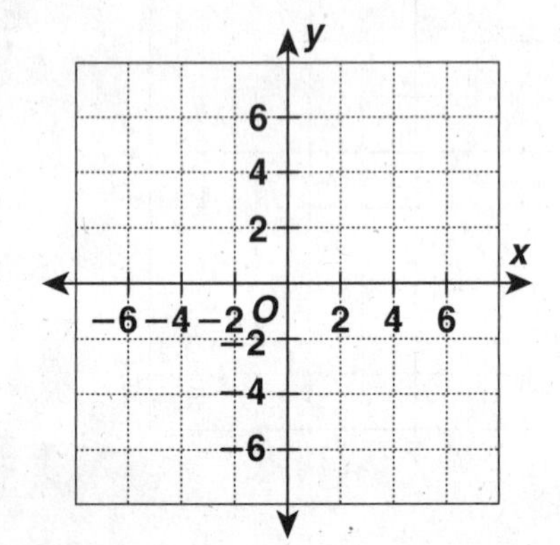

slope is: ______________

2. (−2, −5) and (−2, 2)

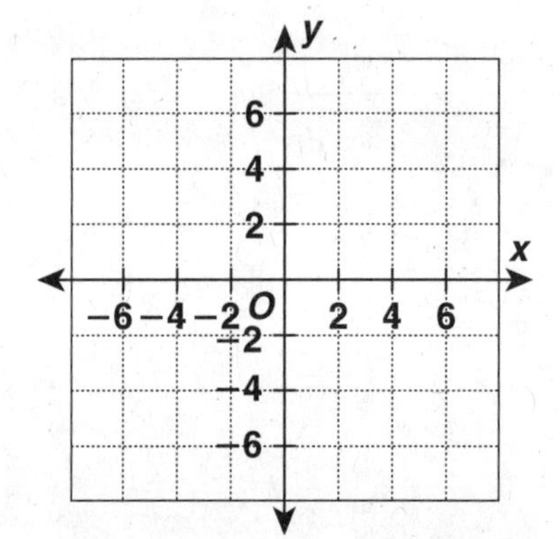

slope is: ______________

3. (1, 2) and (5, −2)

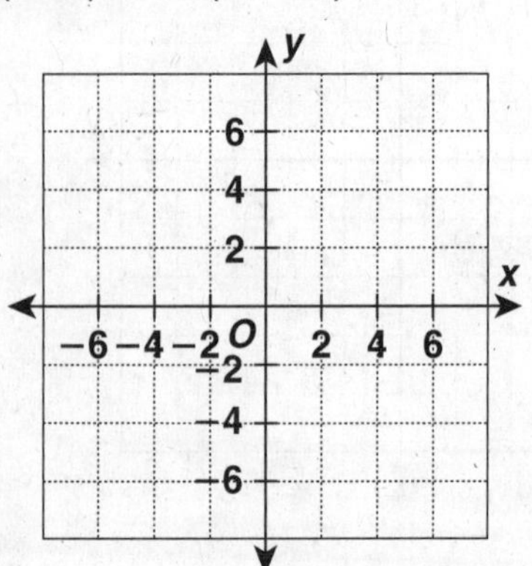

slope is: ______________

4. (−2, −2) and (4, −2)

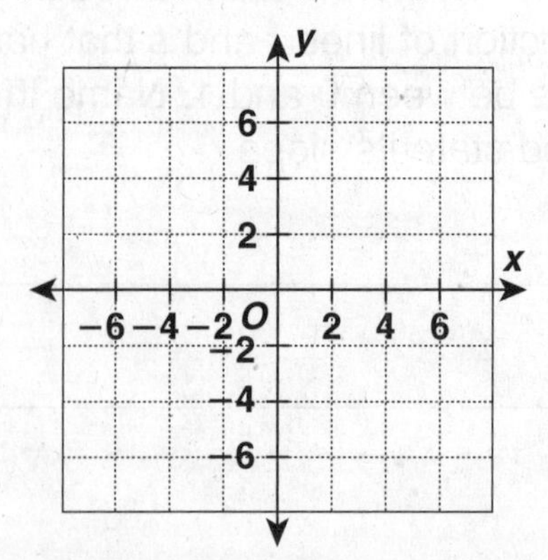

slope is: ______________

Name ______________________ Date __________ Class __________

LESSON 7-5

Reteach

Coordinate Geometry (continued)

To find the slope of a line, use a *direction ratio* such as $\frac{\text{up}}{\text{right}}$.

direction ratio from A to B = $\frac{\text{up } 5}{\text{right } 3}$

slope of $\overleftrightarrow{AB} = \frac{5}{3}$

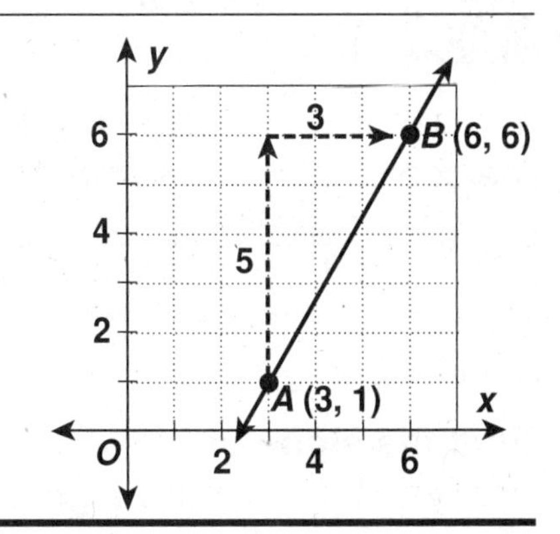

Complete to find the slope of each line.

5.

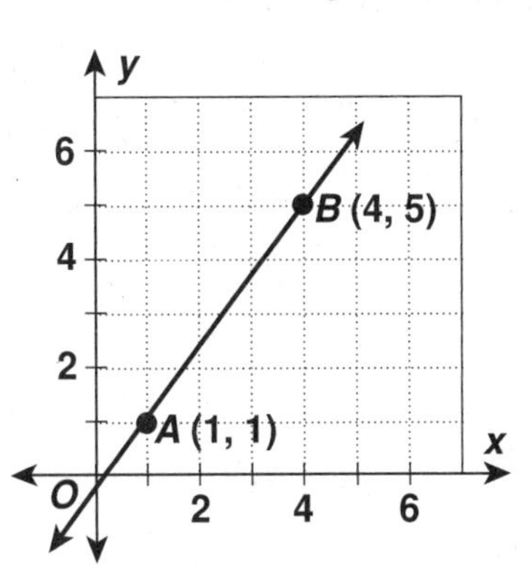

From A to B, do you go up or down? How many units? ________

Do you go right or left? How many units? ____________

slope of $\overleftrightarrow{AB}$ = ____________

The slopes of parallel lines are equal.

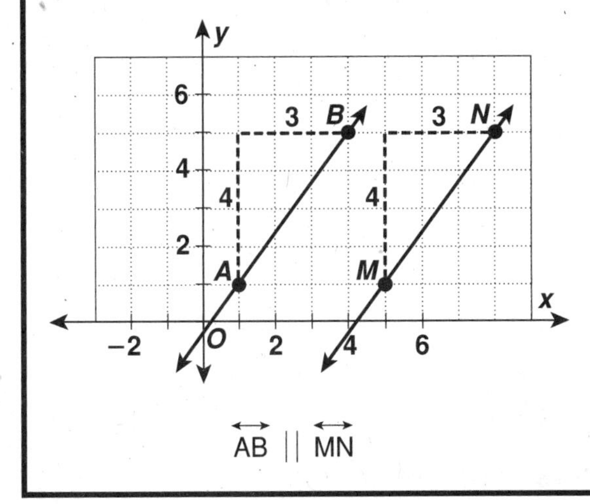

$\overleftrightarrow{AB} \parallel \overleftrightarrow{MN}$

The product of the slopes of perpendicular lines is −1.

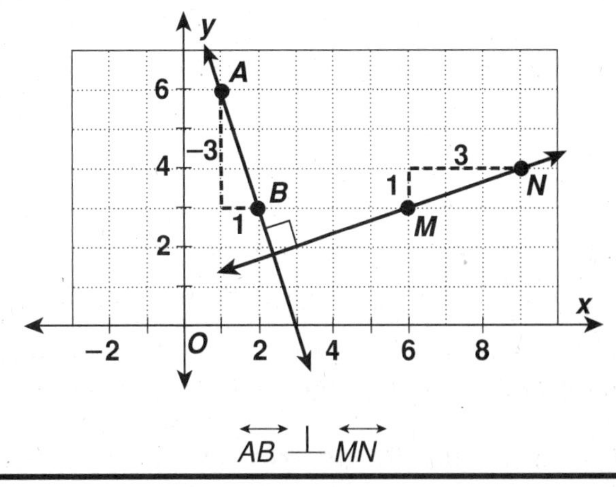

$\overleftrightarrow{AB} \perp \overleftrightarrow{MN}$

Complete each statement. If the slope of $\overleftrightarrow{CD} = -\frac{2}{3}$

6. and $\overleftrightarrow{CD}$ is parallel to $\overleftrightarrow{XY}$, then the slope of $\overleftrightarrow{XY}$ is: ____________

7. and $\overleftrightarrow{CD}$ is perpendicular to $\overleftrightarrow{PQ}$, then the slope of $\overleftrightarrow{PQ}$ is: ____________

Name ______________________ Date __________ Class __________

LESSON **7-5**

Challenge

Are They Lined Up?

You can find the slope of a line by using the coordinates of two points on the line.

$$\text{slope} = \frac{\text{difference of } y\text{-values}}{\text{difference of } x\text{-values}}$$

Be sure to take the differences in the same order.

To find the slope of $\overleftrightarrow{AB}$ with $A(-2, 5)$ and $B(6, 7)$:

$$\text{slope of } \overleftrightarrow{AB} = \frac{7 - 5}{6 - (-2)} = \frac{2}{8} = \frac{1}{4}, \text{ or}$$

$$\text{slope of } \overleftrightarrow{AB} = \frac{5 - 7}{-2 - 6} = \frac{-2}{-8} = \frac{1}{4}$$

Find the slope of the line joining each pair of points. Verify your result on a graph.

1. (5, 5) and (2, 1) __________

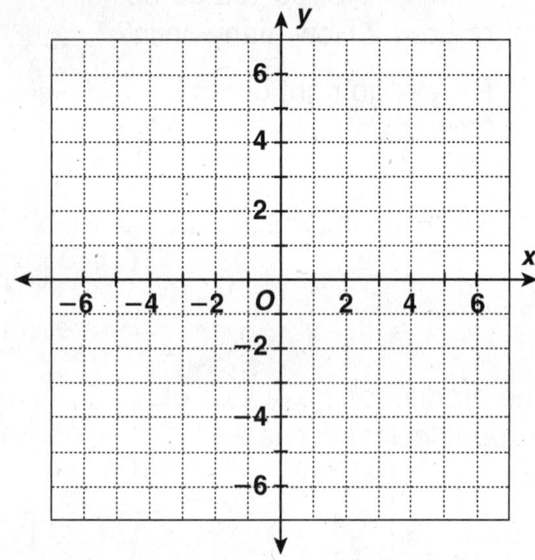

2. (4, 1) and (6, −2) __________

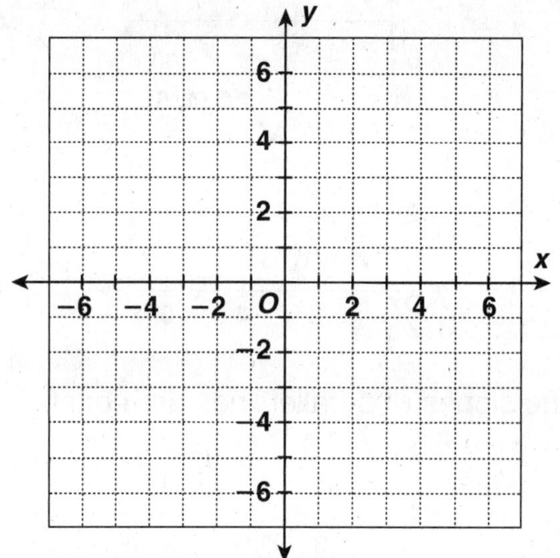

3. Find the value of k so that the slope of the line joining the points $(k, -3)$ and $(4, 2)$ is $\frac{1}{2}$. $k =$ __________

The slope between any two points on a line is the same everywhere on that line; that is, the slope of a given line is *constant*.

Without a graph, determine if each set of points is collinear (lie on the same line). Explain your method.

4. $A(0, -4)$, $B(1, -2)$, and $C(3, 2)$

____

____

5. $P(-7, -1)$, $Q(1, 7)$, and $R(7, 1)$

____

____

6. Find the value of k so that the points $L(-1, 5)$, $M(0, k)$ and $N(1, -1)$ are collinear. $k =$ __________

Name ______________________ Date __________ Class __________

LESSON 7-5

Problem Solving

Coordinate Geometry

The Uniform Federal Accessibility Standards describes the standards for making buildings accessible for the handicapped. The standards say that the least possible slope should be used for a ramp and that the maximum slope of a ramp should be $\frac{1}{12}$.

1. What is the slope of the pictured ramp? Does the ramp meet the standard?

ramp
12 in.
12 ft

____

2. What is the slope of the pictured ramp? Does the ramp meet the standard?

ramp
12 in.
10 ft

____

Write the correct answer.

3. Find the slope of the roof.

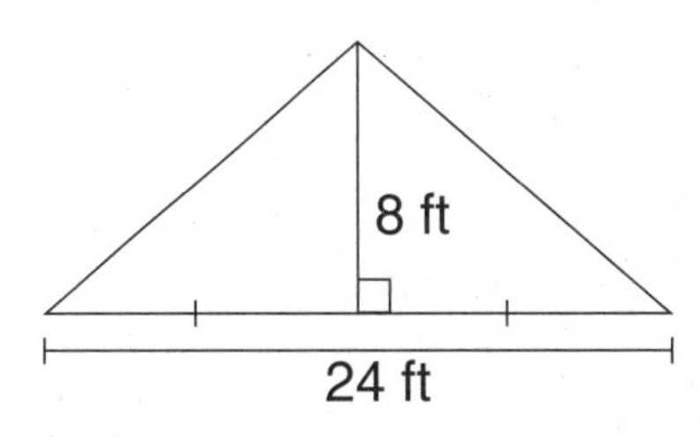

____

Choose the letter that represents the slope.

4. Many building codes require that a staircase be built with a maximum rise of 8.25 inches for a minimum tread width (run) of 9 inches.

A $\frac{8}{9}$ **C** $\frac{9}{8.25}$

B $\frac{11}{12}$ **D** $\frac{12}{11}$

5. Hills that have a rise of about 10 feet for every 17 feet horizontally are too steep for most cars.

F $\frac{10}{17}$ **H** $\frac{17}{10}$

G $\frac{2}{5}$ **J** $\frac{3}{5}$

6. At its steepest part, an intermediate ski run has a rise of about 4 feet for 10 feet horizontally.

A $\frac{2}{5}$ **C** $\frac{5}{2}$

B $\frac{4}{5}$ **D** $\frac{5}{4}$

7. Black diamond, or expert, ski slopes often have a rise of 10 feet for every 14 feet horizontally.

F $\frac{7}{5}$ **H** $\frac{5}{7}$

G $\frac{2}{7}$ **J** $\frac{7}{2}$

Name ______________________ Date ____________ Class ____________

LESSON 7-5

Reading Strategies

Using Graphic Aids

When a horizontal number line crosses a vertical number line, a **coordinate plane,** or grid, is formed. The point where the two lines meet is called the **origin,** or (0, 0).

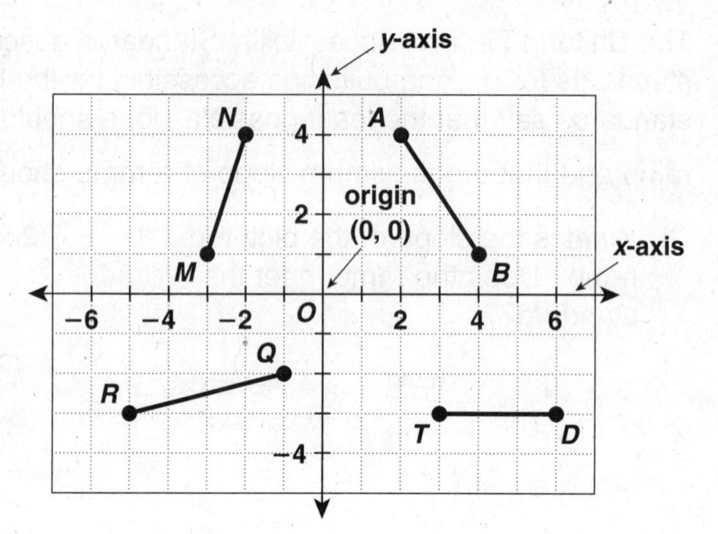

Any location on the coordinate plane can be shown by a point. The pair of numbers that name a point on the coordinate plane is called an **ordered pair.**

Two numbers are needed to identify the location of a point in a coordinate plane.

- The first number tells how far to the left or right of the origin the point lies.
- The second number tells how far up or down from the origin the point lies.

The location of point *M*, starting at (0, 0), is (–3, 1) ⟶ left 3, up 1.

1. Name the ordered pair for point *N*.

____

2. What ordered pair names point *D*?

____

The **slope** is the slant of a line or line segment. Some slopes are steeper than others. Segment *TD* has no slope.

3. Which line segment on the coordinate plane looks like it has the steepest slope?

____

4. Of line segments *MN*, *AB*, and *RQ*, which has the least amount of slope?

____

Name ______________________ Date __________ Class __________

LESSON **7-5**

Puzzles, Twisters & Teasers

How Coordinated Are You?

Determine the slope of each line. In the letter box find the letter that matches each slope. Use the letter to fill in the blanks and solve the riddle.

1. ______________

2. ______________

3. ______________

4. ______________

5. ______________

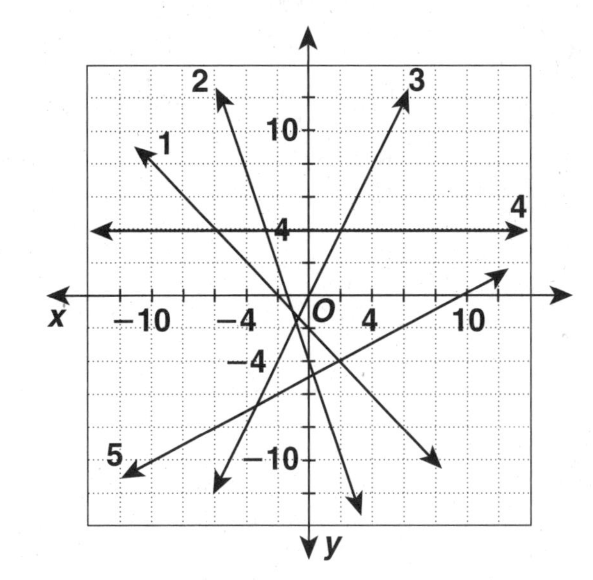

Letter Box
A = $\frac{1}{4}$
D = $-\frac{1}{4}$
E = undefined
I = −1
L = $\frac{1}{2}$
P = 2
R = 1
S = 0
T = $\frac{1}{3}$
U = −3
W = 4

Why did the teacher go to the eye doctor?

She couldn't control her

_____ _____ _____ _____ _____ _____.

2   −3   2   −1 $\frac{1}{2}$ 0

Name ______________________________ Date __________ Class __________

LESSON 7-6

Practice A

Congruence

Match each polygon in column A with a congruent polygon in column B.

Column A		Column B
1. 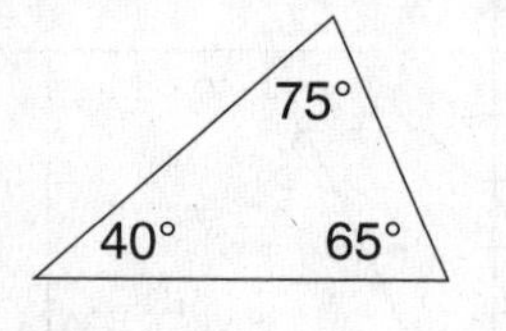	______	A. 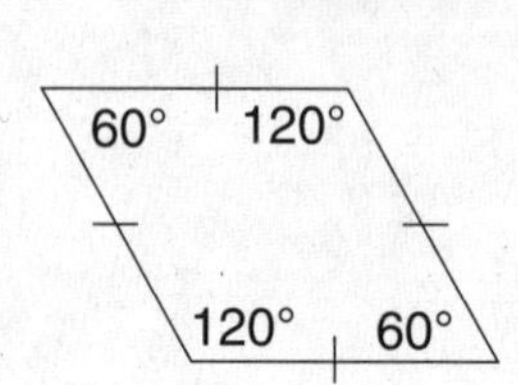
2. 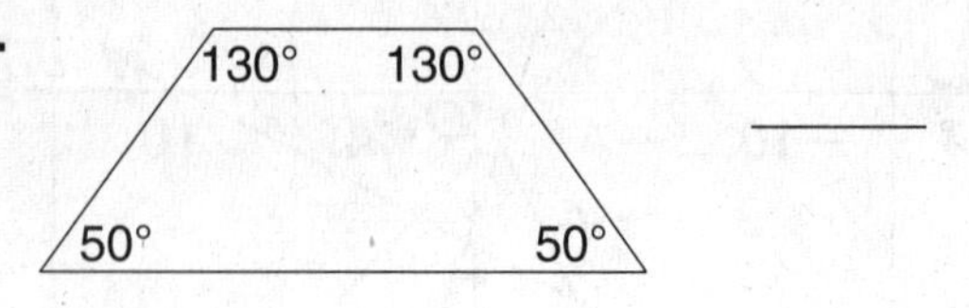	______	B. 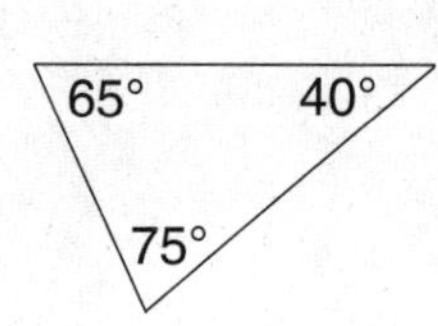
3. 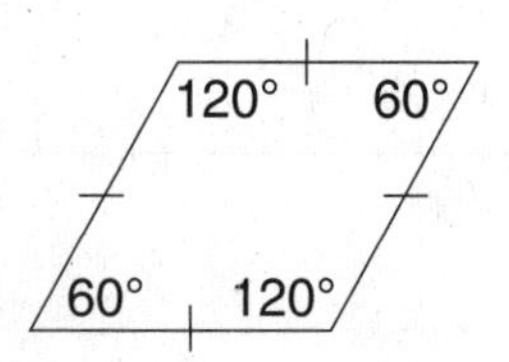	______	C. 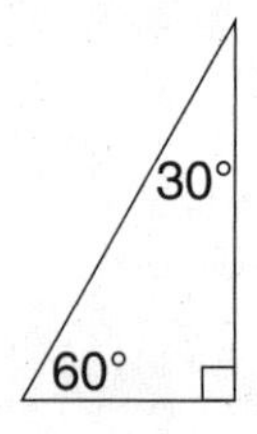
4. 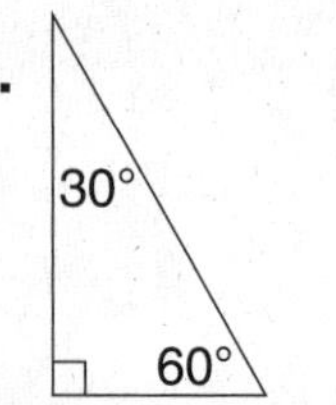	______	D. 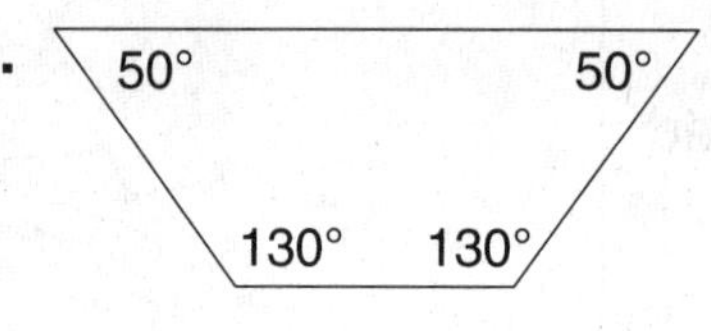

Write a congruence statement.

5. 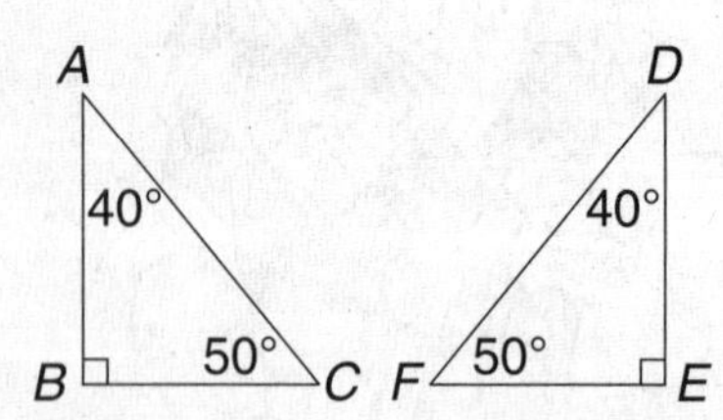

____

____

6. 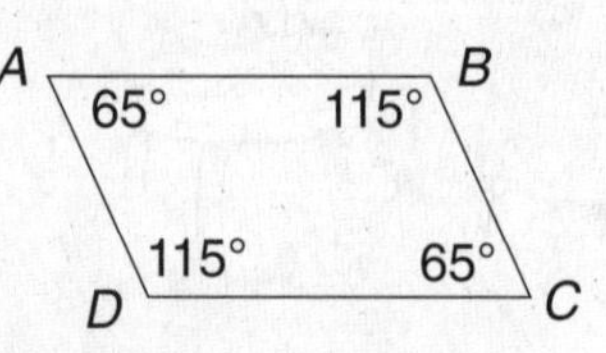

____

____

Triangle *ABC* is congruent to triangle *WXY*.

7. Find d. ______

8. Find t. ______

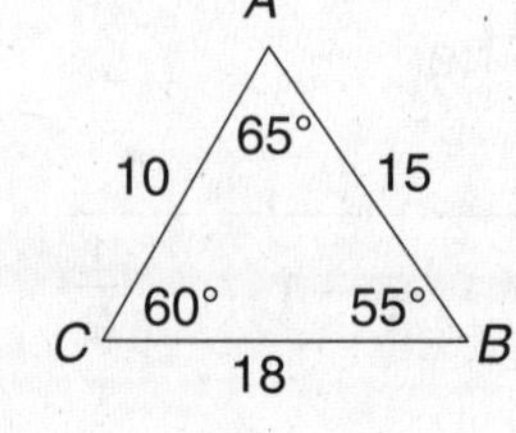

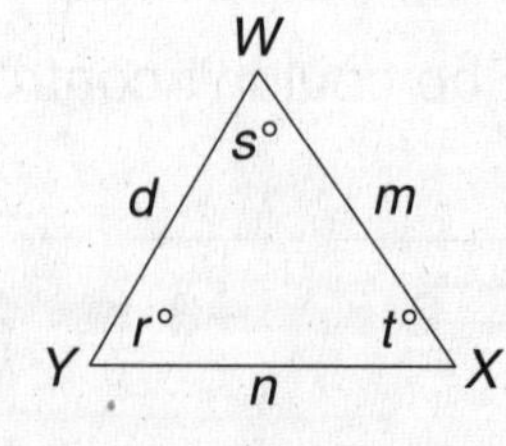

Name ______________________ Date __________ Class __________

LESSON 7-6
Practice B
Congruence

Write a congruence statement for each pair of polygons.

1.

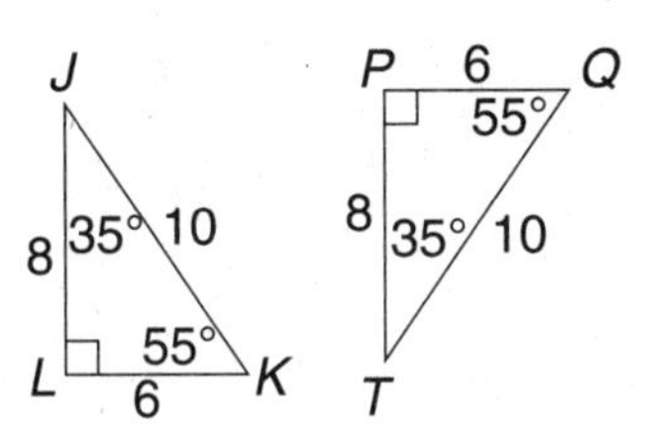

____

____

2.

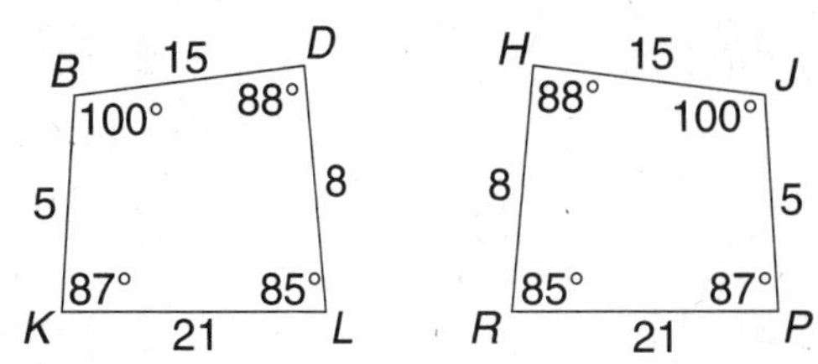

____

____

3.

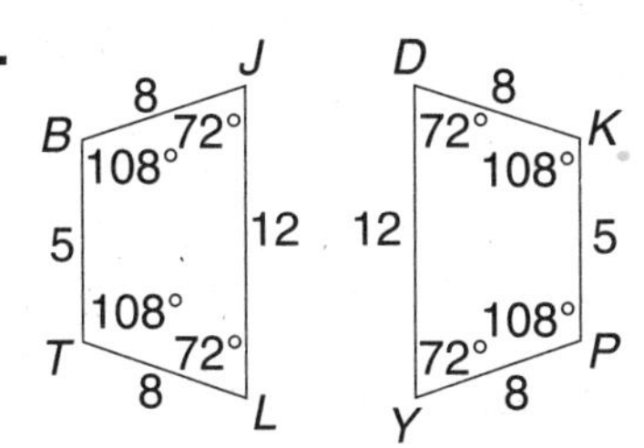

____

____

4.

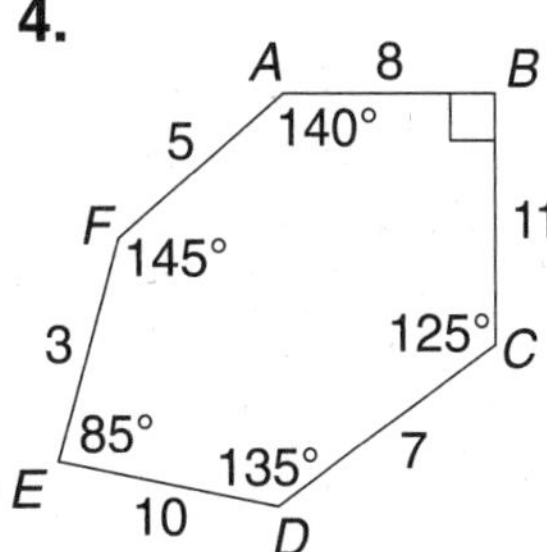

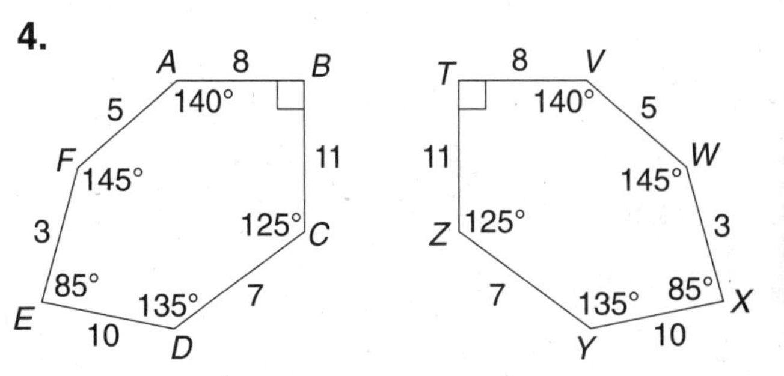

____

____

In the figure, triangle *PRT* ≅ triangle *FJH*.

5. Find *a*. ____________

6. Find *b*. ____________

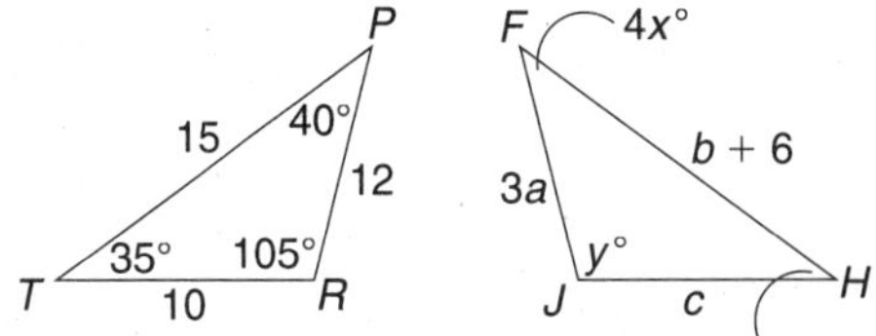

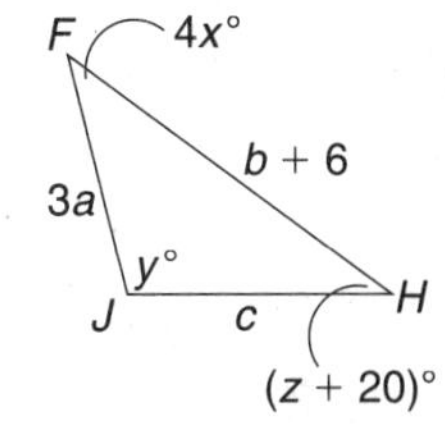

7. Find *c*. ____________

8. Find *x*. ____________

9. Find *y*. ____________

10. Find *z*. ____________

Name ______________________ Date ____________ Class ____________

LESSON 7-6

Practice C

Congruence

In the figure, quadrilateral *ABCD* ≅ quadrilateral *YZWX*.

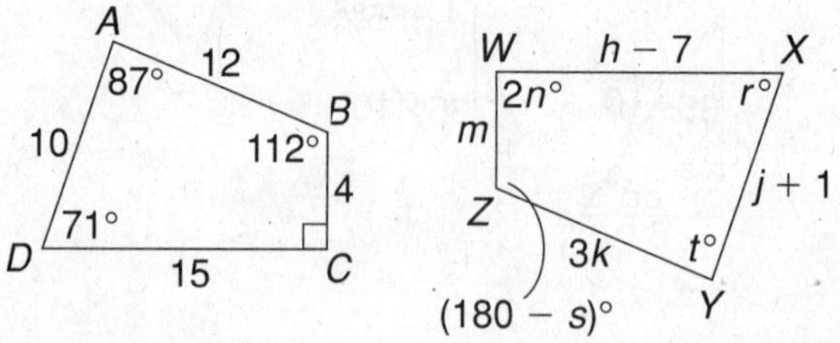

1. Find *m*. ____________

2. Find *h*. ____________

3. Find *j*. ____________

4. Find *k*. ____________

5. Find *n*. ____________

6. Find *s*. ____________

7. Find *t*. ____________

8. Find *r*. ____________

Find the value of the variables if octagon *ABCDEFGH* is congruent to octagon *VWXYZSTU*.

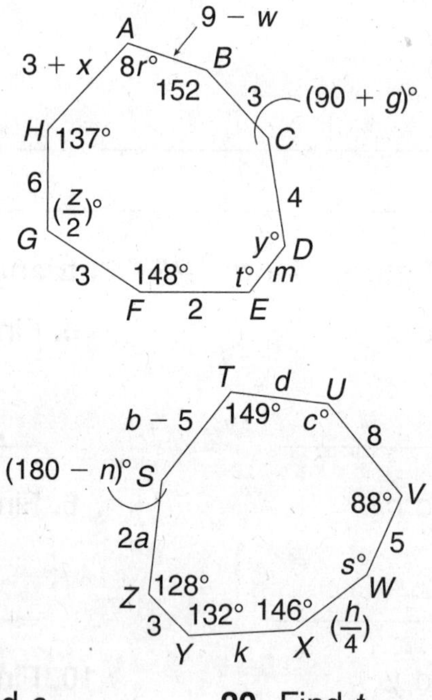

9. Find *a*. ____________

10. Find *b*. ____________

11. Find *c*. ____________

12. Find *d*. ____________

13. Find *g*. ____________

14. Find *h*. ____________

15. Find *k*. ____________

16. Find *m*. ____________

17. Find *n*. ____________

18. Find *r*. ____________

19. Find *s*. ____________

20. Find *t*. ____________

21. Find *w*. ____________

22. Find *x*. ____________

23. Find *y*. ____________

24. Find *z*. ____________

Name ______________________ Date __________ Class __________

LESSON 7-6

Reteach

Congruence

Congruent polygons have the same size and shape.

Corresponding angles are congruent.

$\angle J \cong \angle J'$ $\angle K \cong \angle K'$ $\angle L \cong \angle L'$

(Read *J'* as *J prime.*)

Corresponding sides are congruent.

$\overline{JK} \cong \overline{J'K'}$ $\overline{KL} \cong \overline{K'L'}$ $\overline{LJ} \cong \overline{L'J'}$

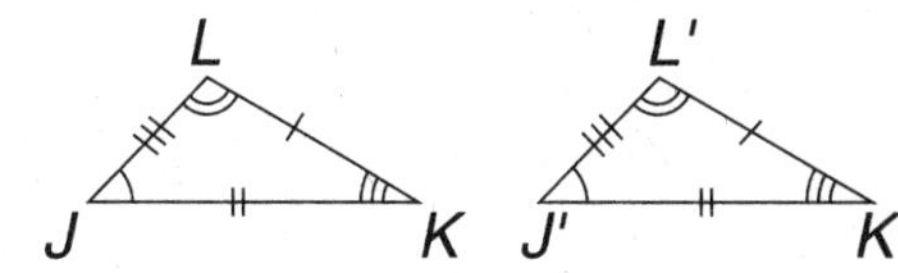

In a congruence statement, the vertices of the second polygon are written in order of correspondence with the first polygon. $\triangle JKL \cong \triangle J'K'L'$

Use the markings in each diagram.
Complete to write each congruence statement.

1.

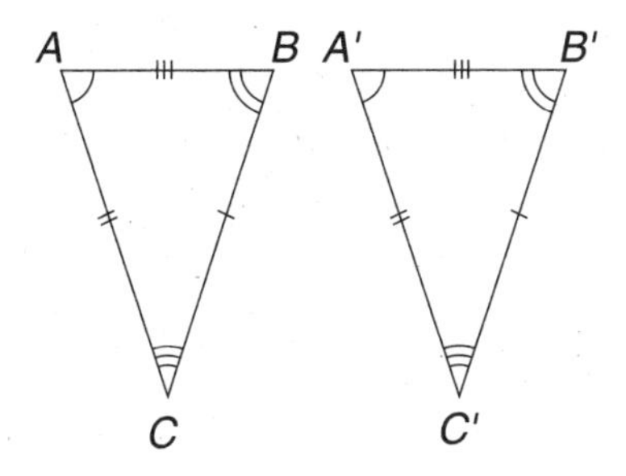

$\angle A \cong$ ______, $\angle B \cong$ ______,

$\angle C \cong$ ______, $\overline{AB} \cong$ ______,

$\overline{BC} \cong$ ______, $\overline{AC} \cong$ ______,

$\triangle ABC \cong$ __________

2.

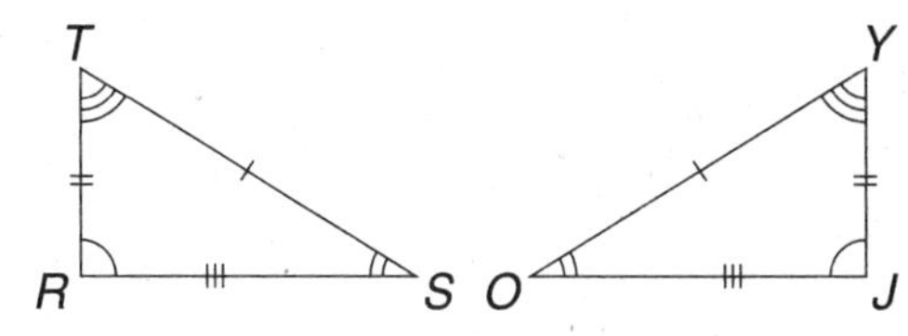

$\angle R \cong$ ______, $\angle S \cong$ ______,

$\angle T \cong$ ______, $\overline{RS} \cong$ ______,

$\overline{RT} \cong$ ______, $\overline{TS} \cong$ ______,

$\triangle RST \cong$ __________

3.

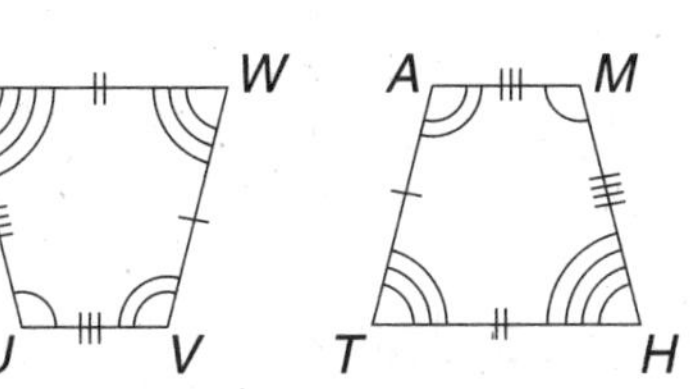

$\angle U \cong$ ______, $\angle V \cong$ ______, $\angle W \cong$ ______, $\angle X \cong$ ______,

$\overline{UV} \cong$ ______, $\overline{VW} \cong$ ______, $\overline{WX} \cong$ ______, $\overline{XU} \cong$ ______

quad. $UVWX \cong$ ______________

Name ______________________ Date __________ Class __________

LESSON 7-6

Reteach

Congruence (continued)

Congruence relations can be used to find unknown values.

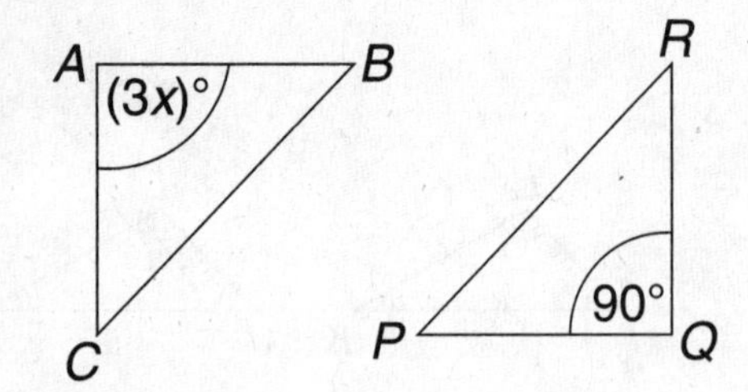

$\angle A \cong \angle Q$

$3x = 90$

$\frac{3x}{3} = \frac{90}{3}$

$x = 30$

$\triangle ABC \cong \triangle QPR$

Using the congruence relationship, complete to find each unknown value.

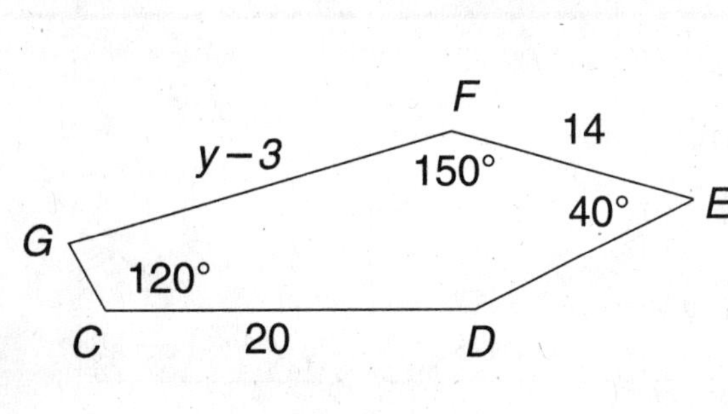

S
R
(2t)°
M
110°
10
$\left(\frac{z}{2}\right)°$
A
Z
3x + 2

4. $\overline{GF} \cong \overline{RA}$

$y - 3 =$ ______

______ ______

$y =$ ______

5. $\angle C \cong$ ______

______ $= 2t$

______ $= \frac{2t}{}$

______ $= t$

6. $\angle A \cong$ ______

$\frac{z}{2} =$ ______

______ $\times \frac{z}{2} =$ ______ $\times$ ______

$z =$ ______

7. $\overline{AZ} \cong$ ______

$3x + 2 =$ ______

______ ______

$3x =$ ______

$\frac{3x}{} =$ ______

$x =$ ______

Name ______________________________ Date __________ Class __________

Challenge
Cloning

In the following exercises, you will construct a triangle congruent to $\triangle ABC$ by copying three strategic parts.

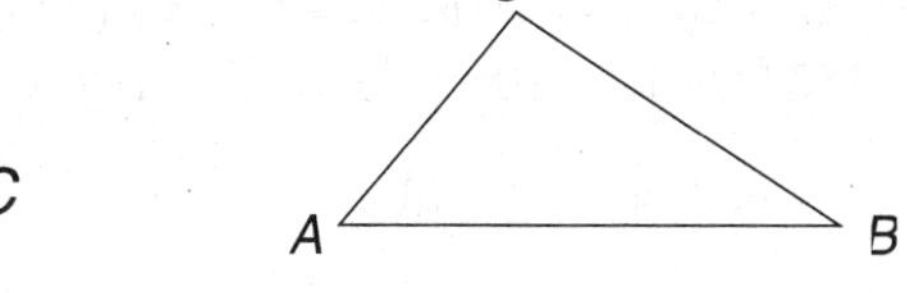

1. To construct a triangle congruent to $\triangle ABC$ by using two pairs of sides and the included angle:

a. With your compass, measure AB.
Copy AB onto line ℓ.
Call the copied length $A'B'$
(read *A prime B prime*).

b. With your compass, measure $\angle A$.
Copy $\angle A$ at vertex A'
with one side as $\overline{A'B'}$.

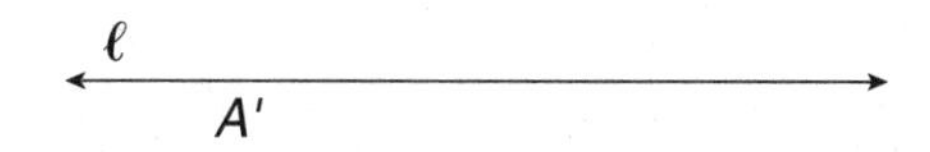

c. With your compass, measure AC.
Copy AC onto the other side of $\angle A'$,
beginning at A' and ending at C'.

d. Draw $\overline{C'B'}$. Use a ruler and protractor to verify that
$\triangle A'B'C' \cong \triangle ABC$.

2. To construct a triangle congruent to $\triangle ABC$
by using two pairs of angles and the included side:

a. With your compass, measure AB.
Copy AB onto line m.
Call the copied length $A''B''$
(*read A double prime B double prime*).

b. With your compass, measure $\angle A$.
Copy $\angle A$ at vertex A'', with one
side as $\overline{A''B''}$.

c. With your compass, measure $\angle B$.
Copy $\angle B$ at vertex B'', with one
side as $\overline{A''B''}$.

d. Use C'' to label the point where the sides of $\angle A''$ and $\angle B''$ intersect.
Use a ruler and protractor to verify that $\triangle A''B''C'' \cong \triangle ABC$.

Name ______________________ Date ____________ Class ____________

LESSON 7-6

Problem Solving

Congruence

Use the American patchwork quilt block design called Carnival to answer the questions. Triangle $AIH \cong$ Triangle AIB, Triangle $ACJ \cong$ Triangle AGJ, Triangle $GFJ \cong$ Triangle CDJ.

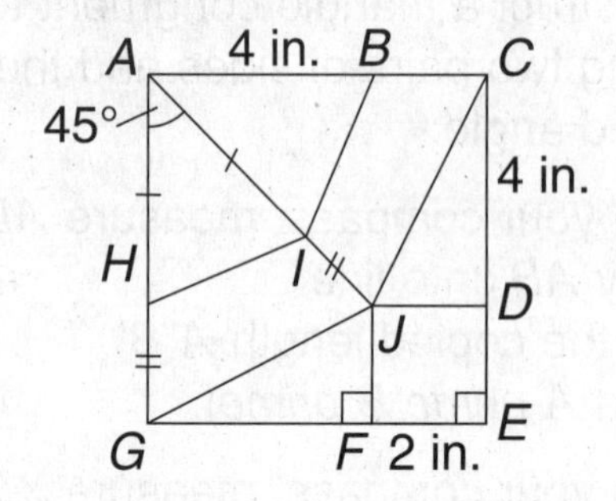

1. What is the measure of $\angle IAB$?

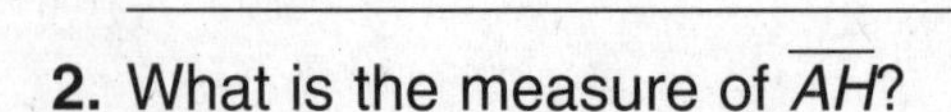

____

2. What is the measure of $\overline{AH}$?

____

3. What is the measure of $\overline{AG}$?

____

4. What is the measure of $\angle JDC$?

____

5. What is the measure of $\overline{FG}$?

____

The sketch is part of a bridge. Trapezoid $ABEF \cong$ Trapezoid $DEBC$. Choose the letter for the best answer.

6. What is the measure of $\overline{DE}$?

A 4 feet
B 8 feet
C 16 feet
D Cannot be determined

7. What is the measure of $\overline{FE}$?

F 4 feet　**H** 8 feet
G 16 feet　**J** 24 feet

8. What is the measure of $\angle FAB$?

A 45°　**C** 60°
B 90°　**D** 120°

9. What is the measure of $\angle ABE$?

F 45°　**H** 60°
G 90°　**J** 120°

10. What is the measure of $\angle EBC$?

A 45°　**C** 60°
B 90°　**D** 120°

11. What is the measure of $\angle BED$?

F 45°　**H** 60°
G 90°　**J** 120°

12. What is the measure of $\angle BCD$?

A 45°　**C** 60°
B 90°　**D** 120°

Name ______________________ Date __________ Class __________

Reading Strategies

Graphic Organizer

This picture helps you understand congruence.

Definition	**Facts**
Two figures that have exactly the same size and the same shape.	• Figures have same size • Figures have same shape
Congruence	
Examples	**Non-examples**

Use the picture to answer these questions.

1. How can you tell if two figures are congruent?

____

2. Is the circle congruent to any other shape? ____________________

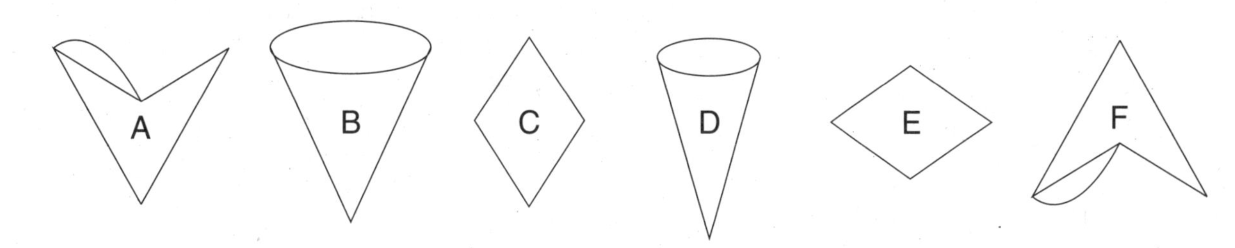

Use the chart and the figures above to answer the following questions.

3. Is figure A congruent to figure B? ____________________

4. Is figure B congruent to figure D? ____________________

5. Is figure C congruent to figure E? ____________________

6. Is figure A congruent to figure F? ____________________

Name ______________________________ Date ____________ Class ____________

LESSON **7-6**

Puzzles, Twisters & Teasers

Equal Time!

In each problem there are congruent figures with three answers. Decide which answer is correct. Write the letter of the correct answer on the blank line which corresponds to the problem.

1. **K** $\triangle KLM \cong \triangle RQS$
R $\angle K \cong \angle Q$
M $\overline{KM} \cong \triangle\overline{QS}$

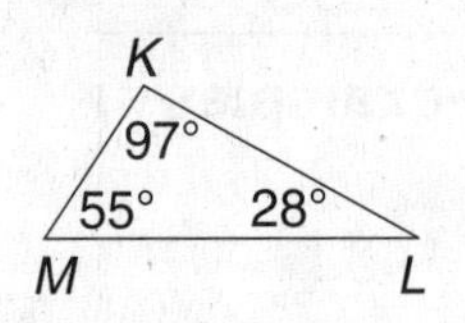

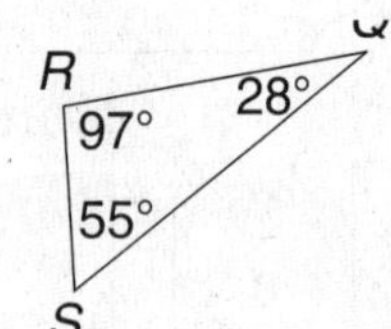

2. **T** $\overline{ED} \cong \overline{JF}$
S $\angle B \cong \angle H$
F pentagon *ABCDE* ≅ pentagon *JIHGF*

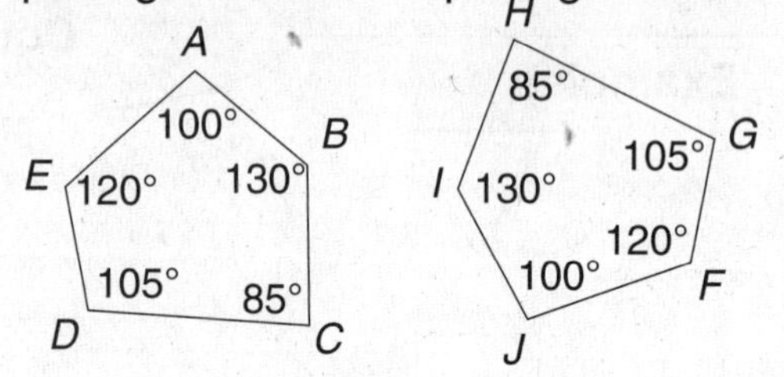

3. **E** $\overline{AB} \cong \overline{FE}$
G $\triangle ACB \cong \triangle DEF$
W $\triangle CAB \cong \triangle EFD$

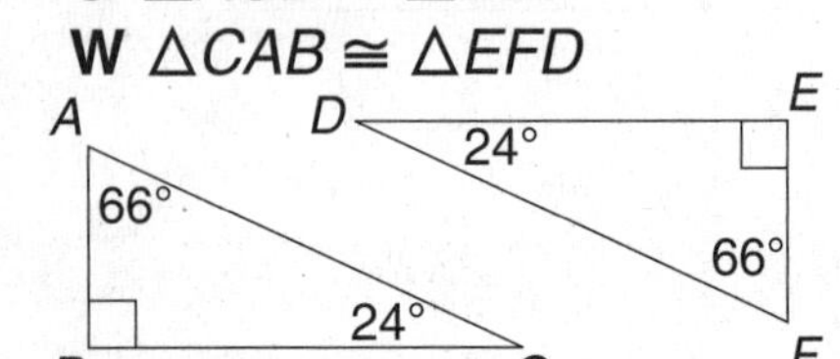

4. **K** quadrilateral *LMNO* ≅ quadrilateral *QRST*
L $\angle N \cong \angle T$
S $\angle L \cong \angle T$

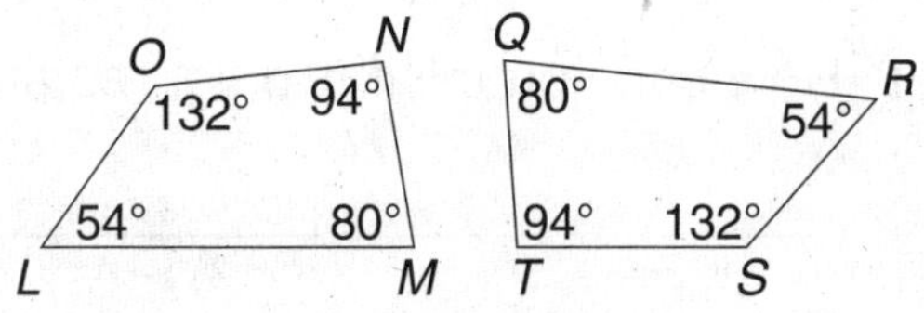

5. **C** $\triangle CAB \cong \triangle JKL$
D $\angle A \cong \angle L$
N $ACB \cong KLJ$

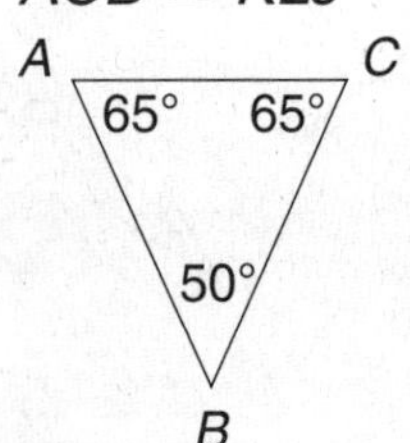

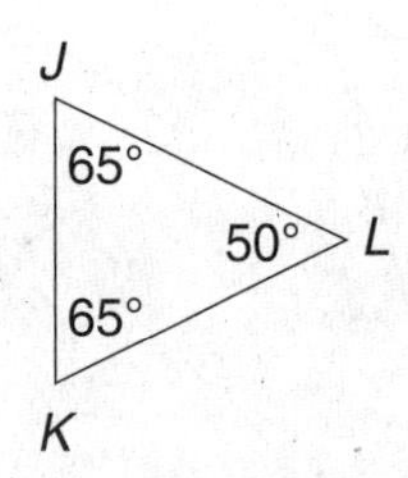

6. **I** quadrilateral *PQSR* ≅ quadrilateral *TUVW*
A $\angle P \cong \angle U$
O $\overline{RP} \cong \overline{TW}$

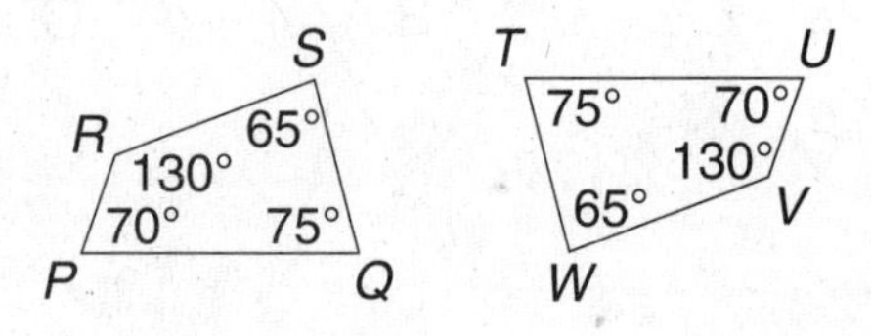

7. **E** $\angle D \cong \angle L$
A $\overline{AD} \cong \overline{ML}$
O $\overline{AB} \cong \overline{NM}$

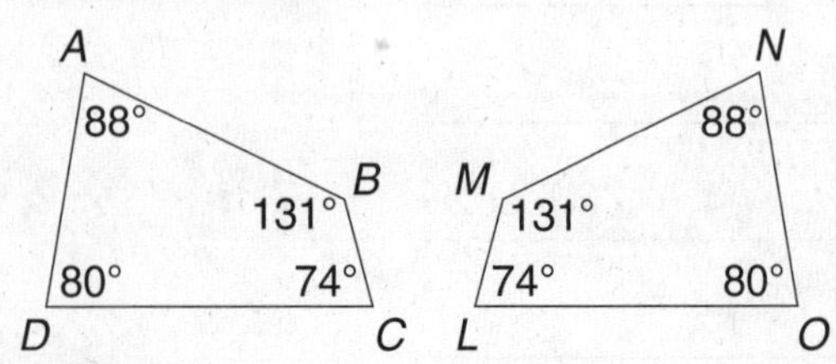

What kind of face has hands but no eyes, nose, or mouth?

A 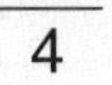   ___ ___   ___ ___ ___ ___

5   4   7   5   1   2   6   5   3

Name ____________________ Date __________ Class __________

LESSON 7-7

Practice A

Transformations

Identify each as a translation, rotation, reflection, or none of these.

1.

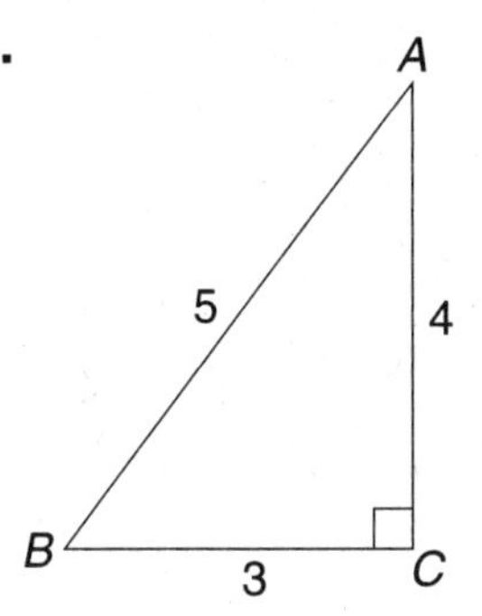

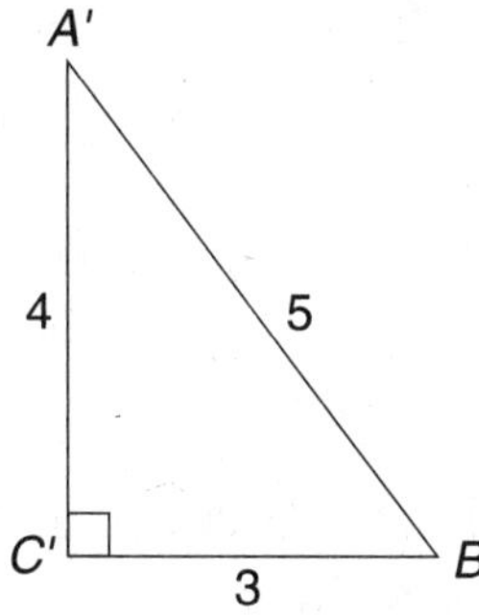

____

2.

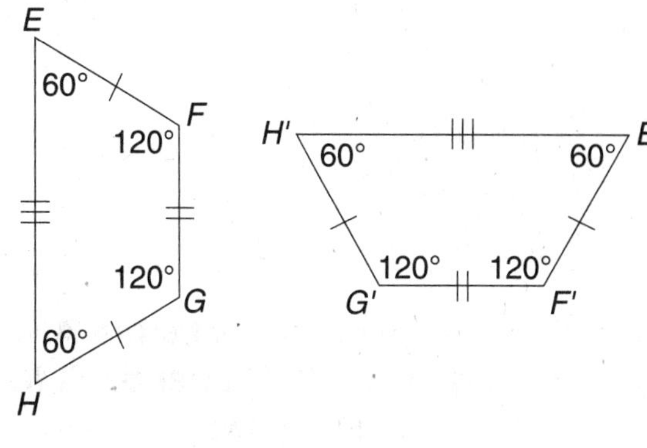

____

3.

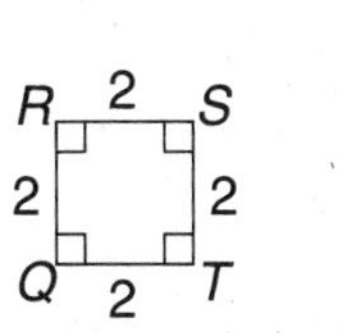

R' 4 S'

4 4

Q' 4 T'

____

4.

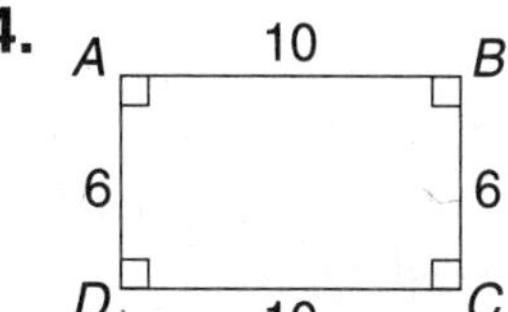

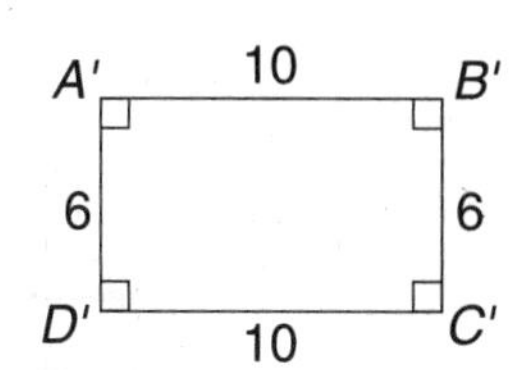

____

5. Draw the image of the triangle ABC with vertices (–2, 2), (2, 4), and (2, 2) after a translation 5 units down.

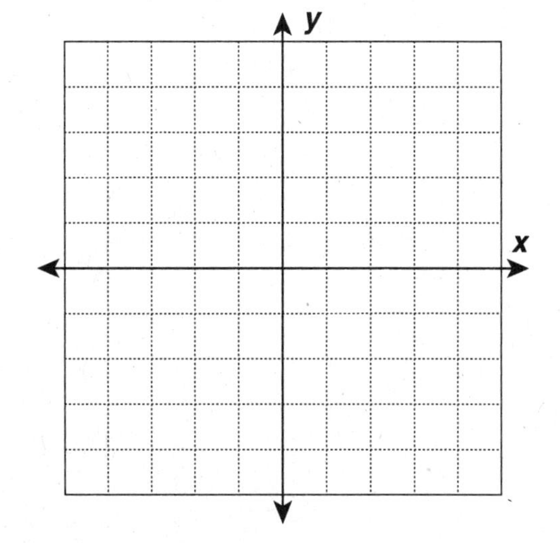

6. Triangle XYZ has vertices X(3, 4), Y(4, 1), and Z(1, 1). Find the coordinates of the image of point Z after a reflection across the y-axis.

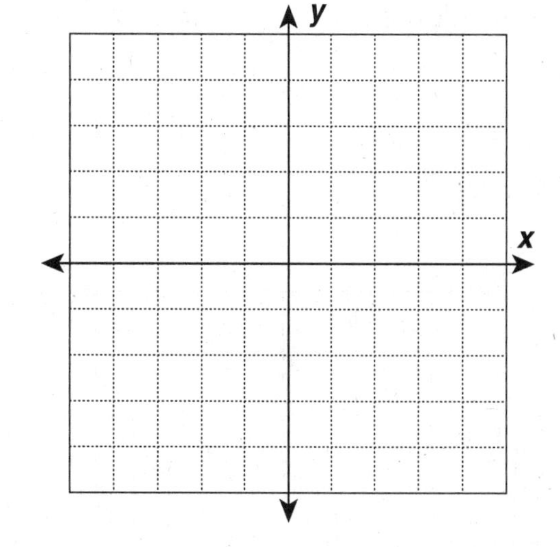

____

Name ______________________ Date ____________ Class ____________

LESSON 7-7

Practice B

Transformations

Identify each as a translation, rotation, reflection, or none of these.

1.

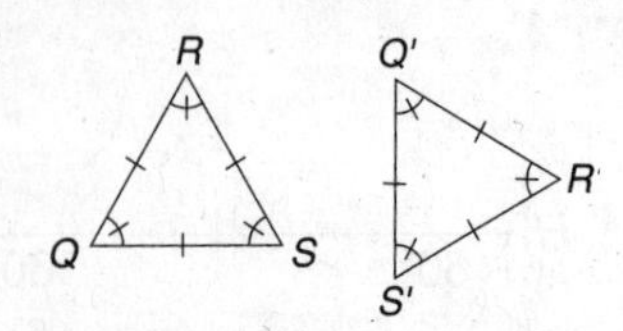

____

2.

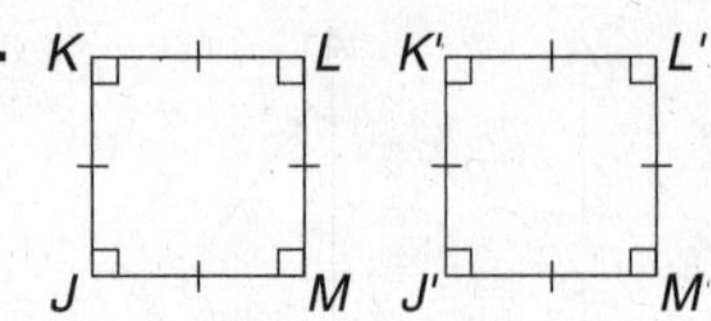

____

Draw the image of the rectangle *ABCD* with vertices (−2, 1), (−1, 3), and (3, 3), (2, 1) after each transformation.

3. translation 3 units down

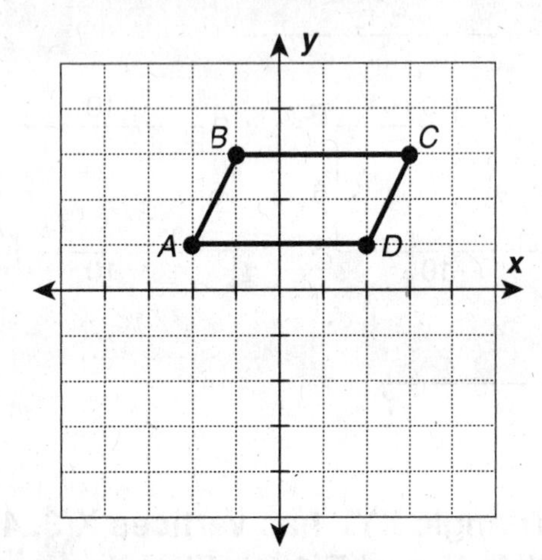

4. 180° rotation around (0, 0)

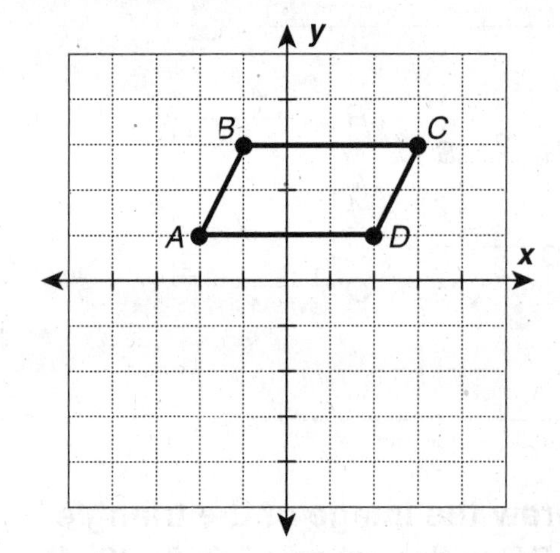

Triangle *ABC* has vertices *A*(−3, 1), *B*(2, 4), and *C*(3, 1). Find the coordinates of the image of each point after each transformation.

5. reflection across the *x*-axis, point *B*

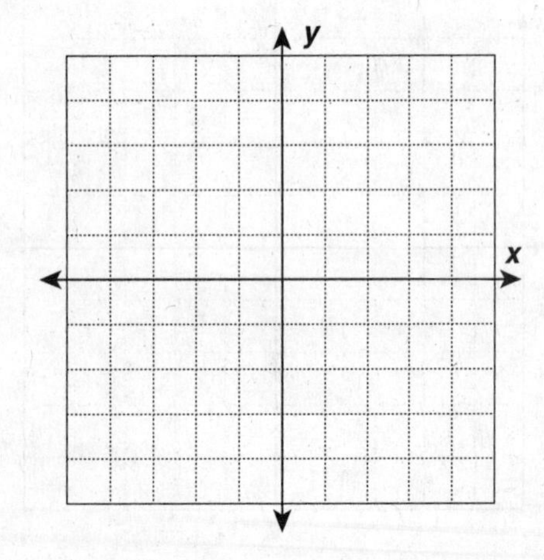

____

6. translation 6 units down, point *A*

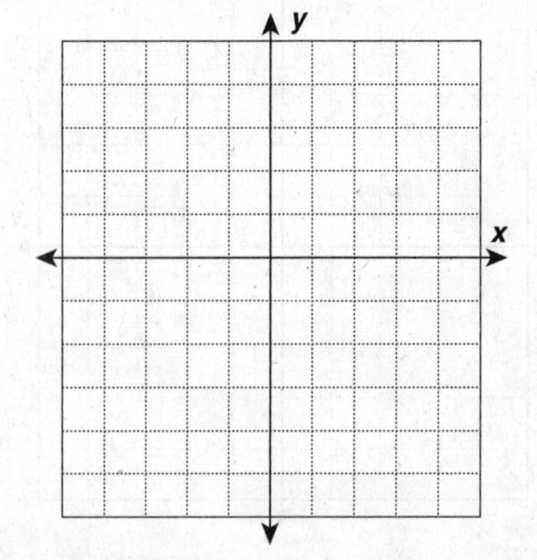

____

Name ______________________ Date __________ Class __________

LESSON 7-7

Practice C

Transformations

Give the coordinates of each point after a reflection across the *x*-axis.

1. (−2, 3) **2.** (−4, −1) **3.** (5, 2) **4.** (6, −3)

__________ __________ __________ __________

Give the coordinates of each point after a reflection across the *y*-axis.

5. (−1, −5) **6.** (3, 2) **7.** (−4, 6) **8.** (7, −2)

__________ __________ __________ __________

Give the coordinates of each point after a 180° rotation around (0, 0).

9. (4, −6) **10.** (−5, 3) **11.** (1, 2) **12.** (−3, −2)

__________ __________ __________ __________

Perform the given transformation.

13. Reflect across line *m*.

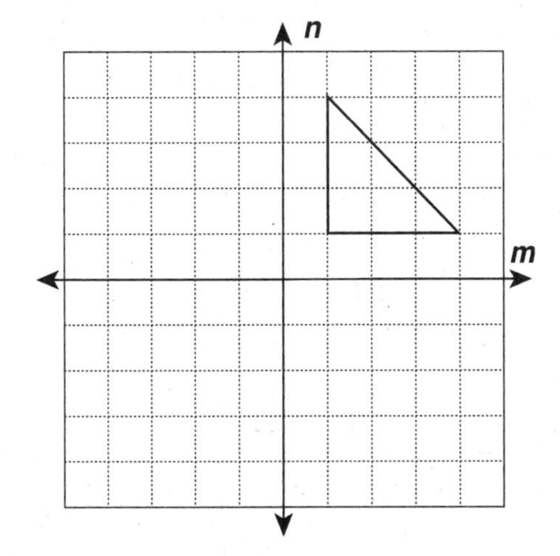

14. Rotate clockwise 180° around (0, 0).

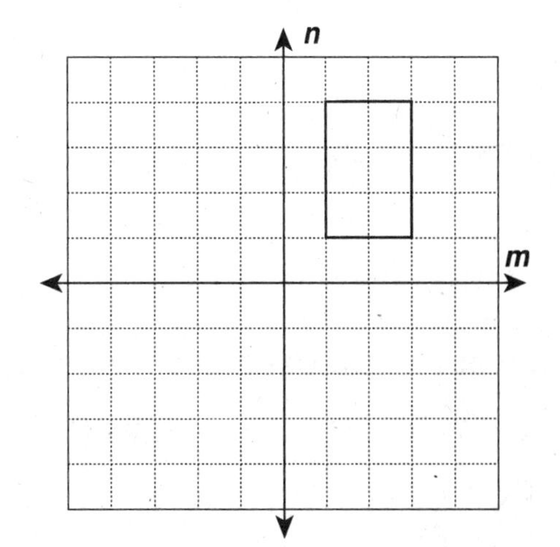

Name ______________________ Date __________ Class __________

LESSON 7-7

Reteach

Transformations

Reflection	Rotation	Translation
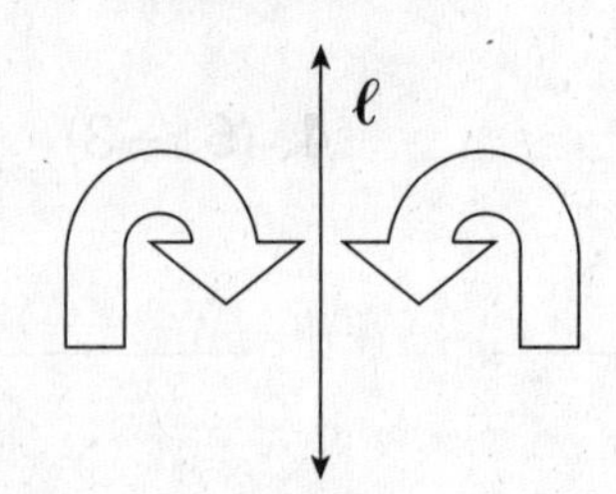	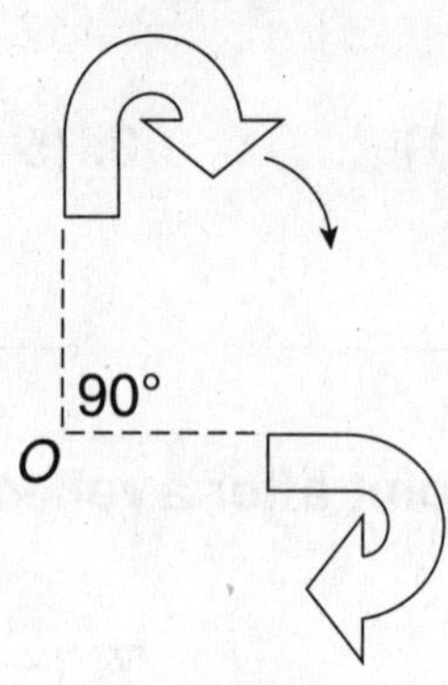	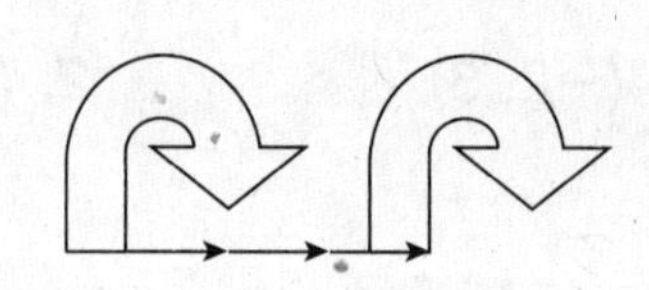
a *mirror image* (a *flip*) The figure is reflected over line ℓ.	a *turning* The figure is rotated 90° clockwise about point *O*.	a *slide* The figure is translated 3 units right and 1 unit up.

Complete to identify each type of transformation.

1.

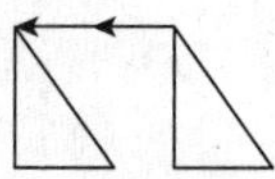

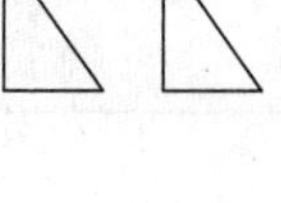

Slide the figure 2 units

____

Transformation:

____

2.

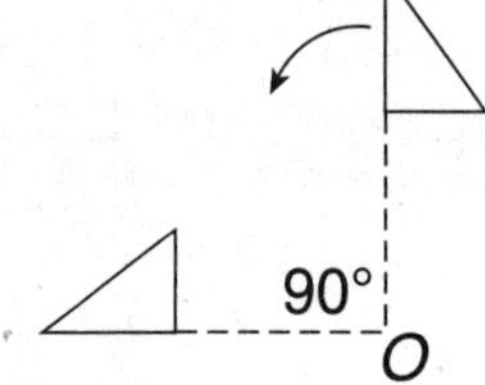

Turn the figure 90°

____

Transformation:

____

3.

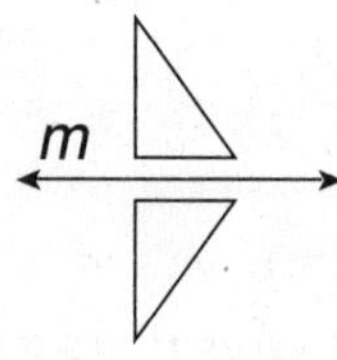

Flip the figure

____

Transformation:

____

Identify each as a translation, rotation, or reflection.

4.

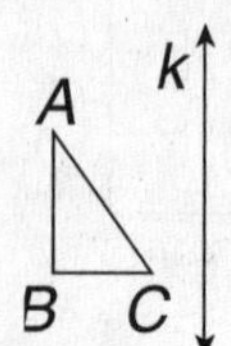

____

5.

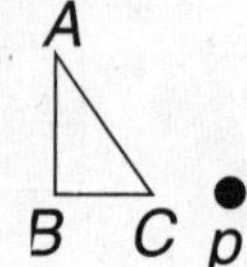

____

6.

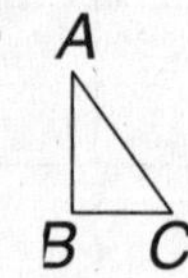

____

Name ______________________ Date __________ Class __________

LESSON 7-7

Reteach

Transformations (continued)

When reflecting a point about a horizontal or vertical line, only one of the coordinates changes.

reflection across *y*-axis
x-coordinate goes to its opposite

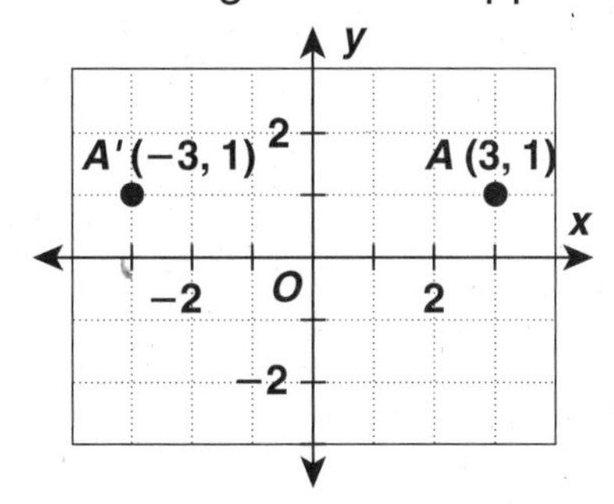

reflection across *x*-axis
y-coordinate goes to its opposite

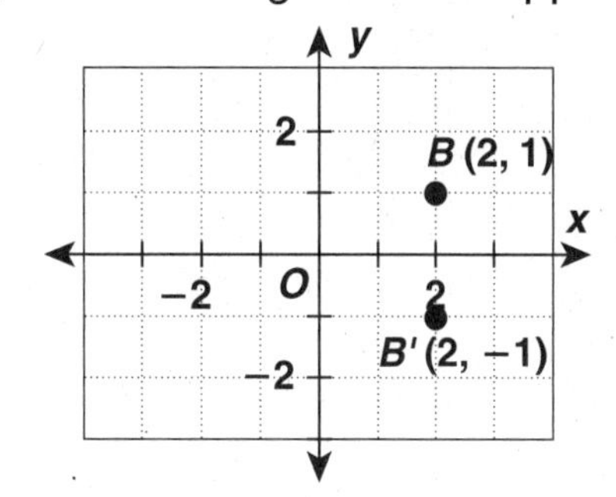

When translating a point, add the indicated number of units to each coordinate.

For a translation left or right, add units to the x-coordinate. For a translation up or down, add units to the y-coordinate.

P(1, 4) is translated 3 units down.

P(1, 4) → P′(1, 4 + (−3)),
or P′(1, 1)

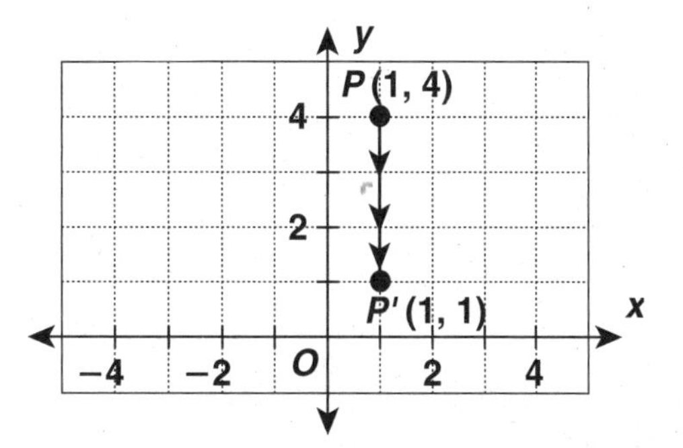

Draw and label the image after the reflection.

7. *P*(−1, 2) over the *y*-axis

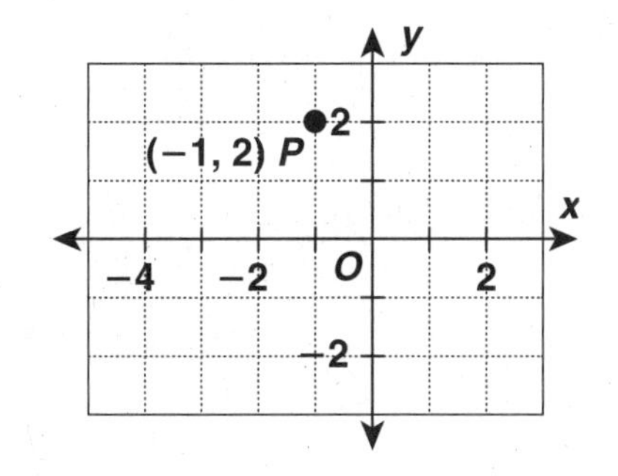

Draw and label the image after the translation.

8. Translate *A*(−3, 5) 4 units to the right.

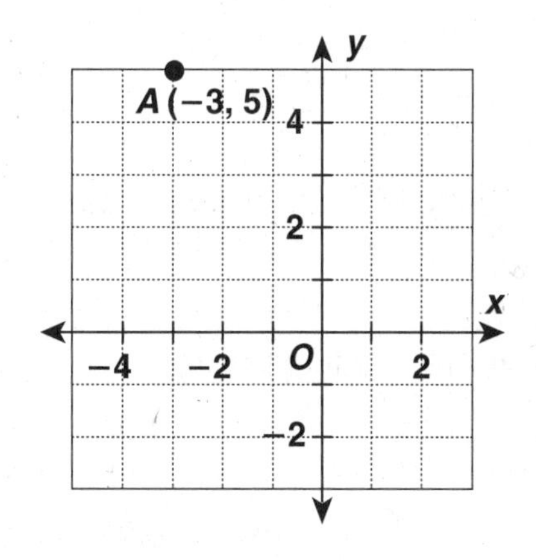

Name ______________________ Date ____________ Class ____________

LESSON **7-7**

Challenge

One Trip Instead of Two

1. Consider $\triangle ABC$ with vertices at $A(4, 1)$, $B(1, 2)$, and $C(3, 5)$.

a. Draw $\triangle A'B'C'$, the image of $\triangle ABC$ after a reflection in the line $x = 1$.

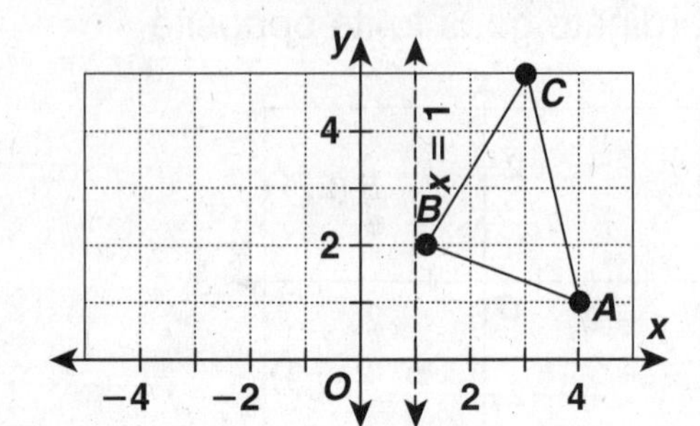

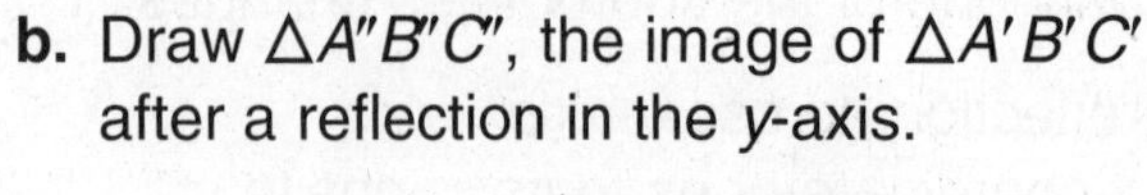

b. Draw $\triangle A''B''C''$, the image of $\triangle A'B'C'$ after a reflection in the y-axis.

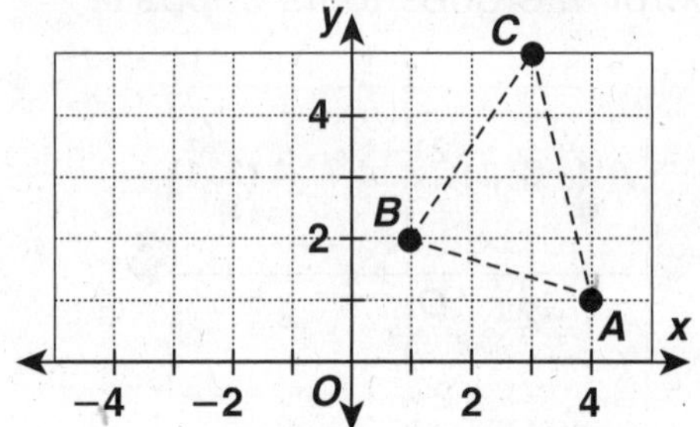

c. Describe a single transformation that takes $\triangle ABC$ to the image $\triangle A''B''C''$.

____

d. How are the lines of reflection related?

____

2. Consider $\triangle PQR$ with vertices at $P(1, 2)$, $Q(3, 4)$, and $R(5, 3)$.

a. Draw $\triangle P'Q'R'$, the image of $\triangle PQR$ after a reflection in the line $y = 1$.

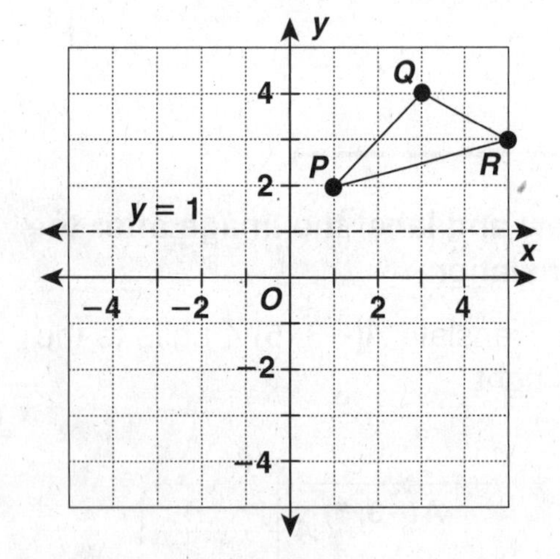

b. Draw $\triangle P''Q''R''$, the image of $\triangle P'Q'R'$ after a reflection in the x-axis.

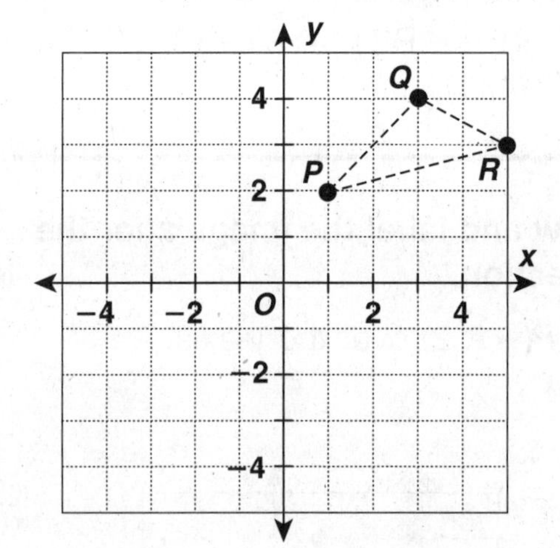

c. Describe a single transformation that takes $\triangle PQR$ to the image $\triangle P''Q''R''$.

____

d. How are the lines of reflection related?

____

Name ______________________ Date __________ Class __________

LESSON 7-7 Problem Solving
Transformations

Parallelogram ABCD has vertices *A*(–3, 1), *B*(–2, 4), *C*(3, 4), and *D*(2, 1). Refer to the parallelogram to write the correct answer.

1. What are the coordinates of point A after a reflection across the x-axis?

____

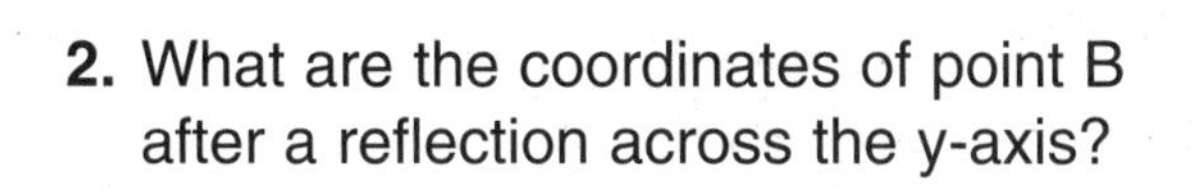

2. What are the coordinates of point B after a reflection across the y-axis?

____

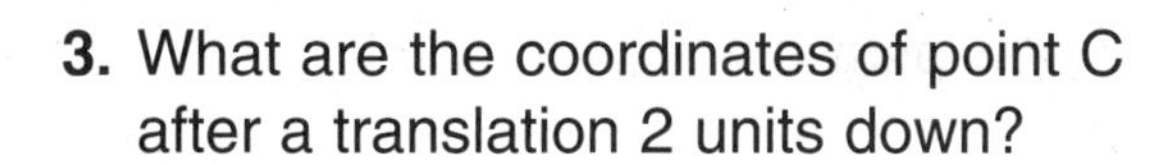

3. What are the coordinates of point C after a translation 2 units down?

____

4. What are the coordinates of point D after a 180° rotation around (0, 0)?

____

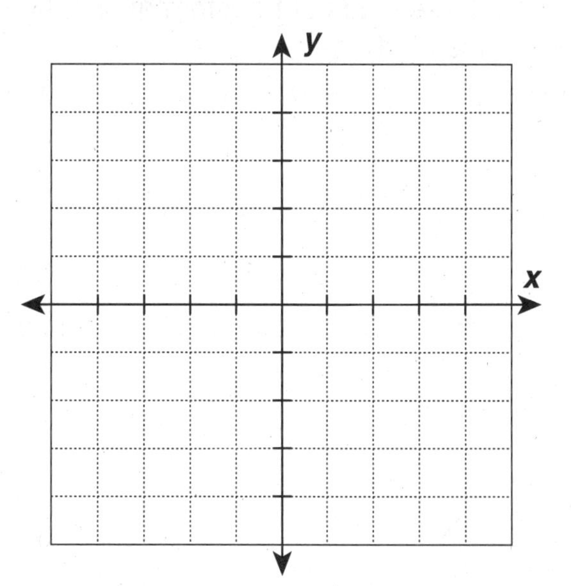

Identify each as a translation, rotation, reflection or none of these.

5.

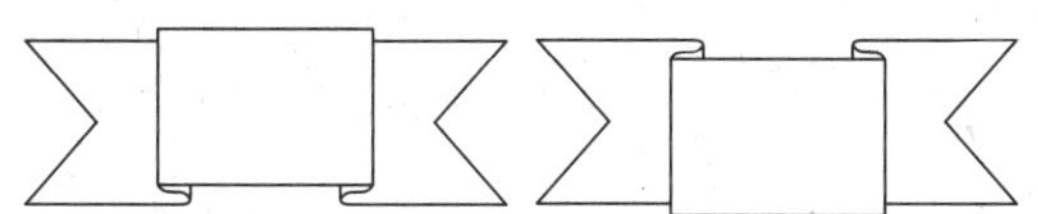

A translation **C** rotation
B reflection **D** none of these

6.

F translation **H** rotation
G reflection **J** none of these

7.

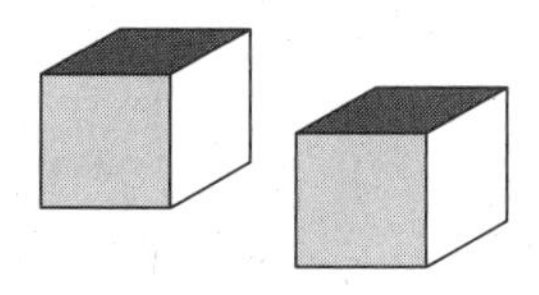

A translation **C** rotation
B reflection **D** none of these

8.

F translation **H** rotation
G reflection **J** none of these

Name ______________________ Date __________ Class __________

LESSON 7-7

Reading Strategies
Focus on Vocabulary

Transformation is a change in the position or size of a figure. The act of transforming the figure does not change its shape.

Here are three kinds of transformations that do not change a figure's size or shape.

- **Translation** ⟶ **Slide** a figure to a new position.

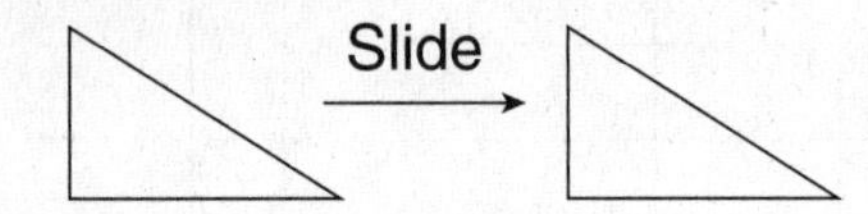

- **Rotation** ⟶ **Turn** a figure around a point to a new position.

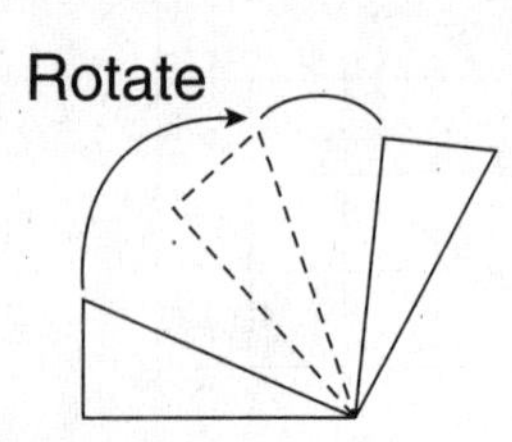

- **Reflection** ⟶ **Flip** a figure over a line for a mirror image.

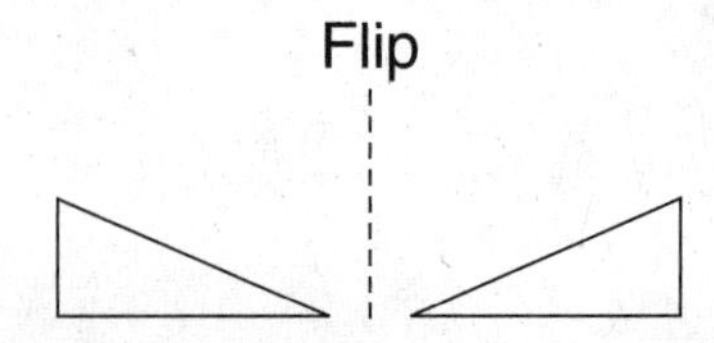

Use *translation, rotation,* or *reflection* to name the transformation shown in each picture.

1.

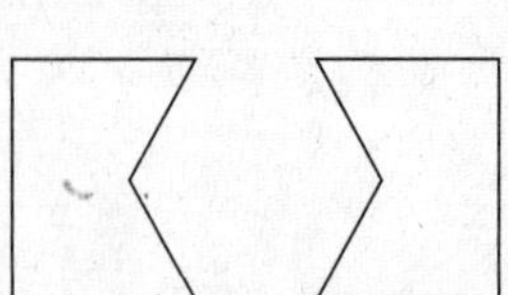

____

2.

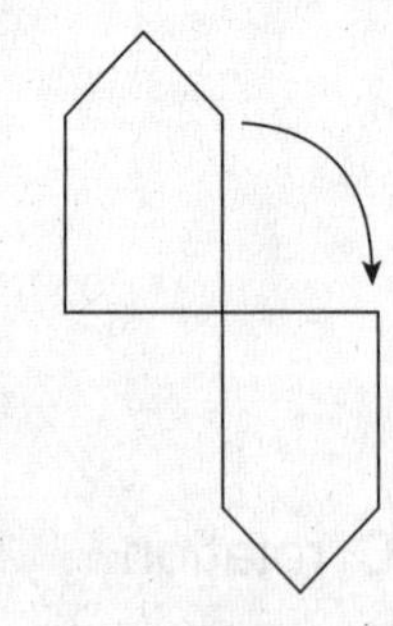

____

Name ________________________________ Date ____________ Class ____________

LESSON **7-7**

Puzzles, Twisters & Teasers

Transformer!

Circle words from the list in the word search (horizontally, vertically or diagonally). Then find a word that answers the riddle. Circle it and write it on the line.

transformation	translation	rotation	center	reflection
image	plane	figure	flip	axis

```
C A B G T D U C K F L I P T
E Q X C D E F G H U R T H R
N O L I U H N T G B O Q W A
T R A N S F O R M A T I O N
E W E R T Y U F E U A I O S
R E F L E C T I O N T P T L
P J G F U Y R G E D I E A A
H L U N E D E U K O O N M T
O I A Y T R E R L D N H J I
P L M N K O I E E R O R F O
I M A G E U Y I O P N E O N
```

What should you yell when you see a low-flying water fowl coming your way?

________________!

Name ______________________ Date ____________ Class ____________

LESSON **7-8**

Practice A

Symmetry

Complete the figure. The dashed line is the line of symmetry.

1.

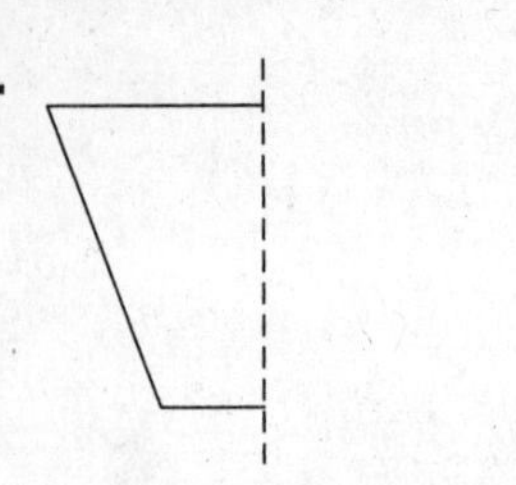

2.

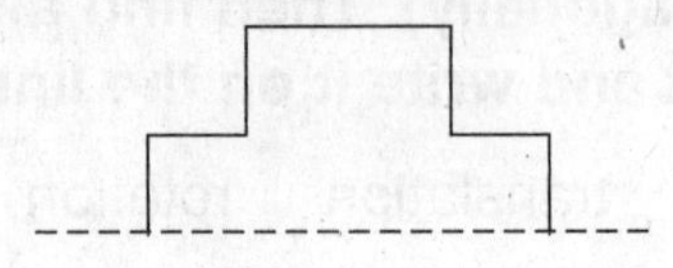

3.

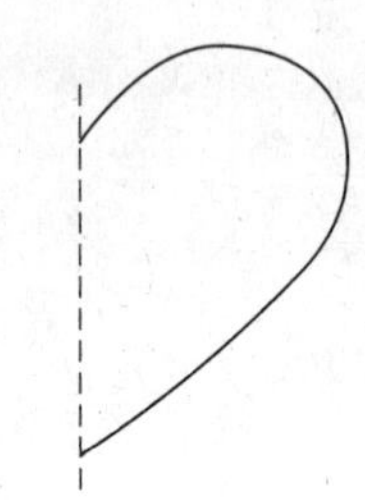

4.

5.

6.

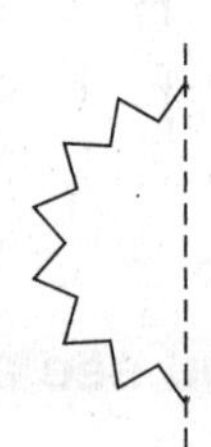

Complete the figure. The point is the center of rotation.

7. 4-fold

8. 2-fold

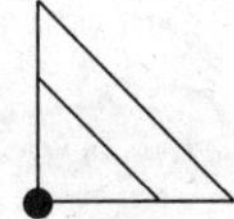

9. 5-fold

10. 3-fold

Name ______________________________ Date ___________ Class ___________

LESSON 7-8

Practice B

Symmetry

Complete each figure. The dashed line is the line of symmetry.

1.

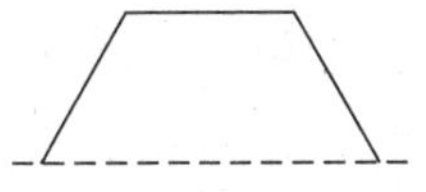

2.

3.

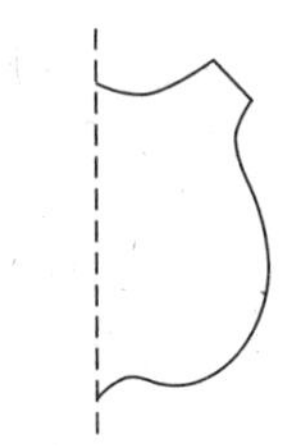

4.

5.

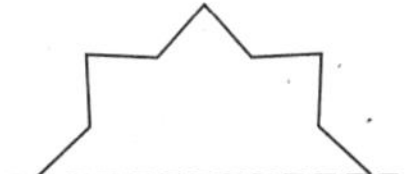

6.

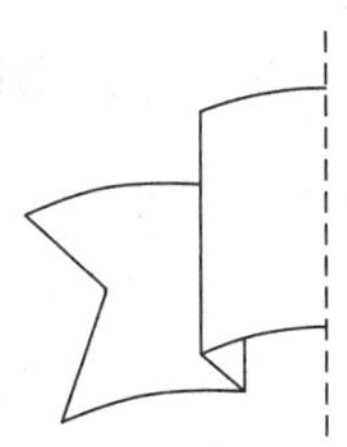

Complete each figure. The point is the center of rotation.

7. 5-fold

8. 4-fold

9. 2-fold

10. 2-fold

Name ______________________ Date ____________ Class ____________

LESSON 7-8

Practice C

Symmetry

Complete each figure. The dashed line is the line of symmetry.

1.

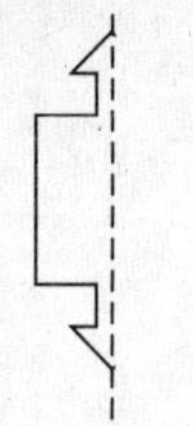

2.

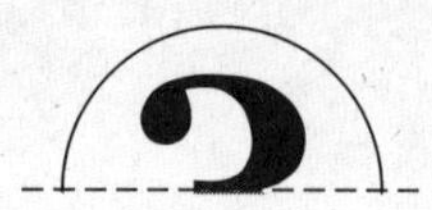

3.

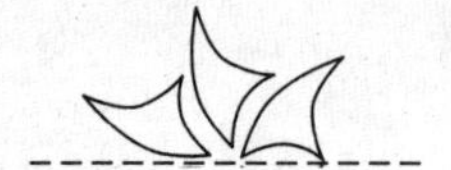

4.

Complete each figure. The point is the center of rotation.

5. 3-fold

6. 5-fold

Draw an example of a figure with each type of symmetry.

7. line symmetry and rotational symmetry

8. line symmetry but not rotational symmetry

Name ______________________ Date __________ Class __________

Reteach
Symmetry

Vertical Line Symmetry | **Horizontal Line Symmetry** | **Diagonal Line Symmetry**

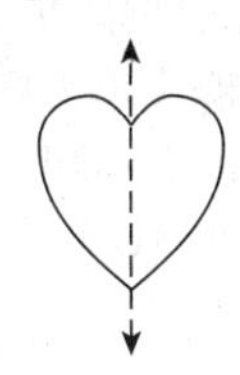

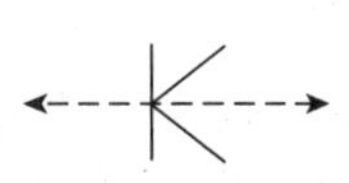

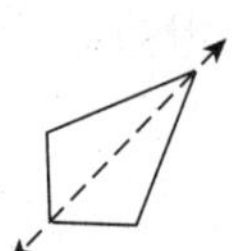

Draw all the lines of symmetry in each figure.

1. 

2. 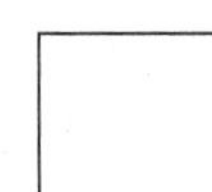

3. 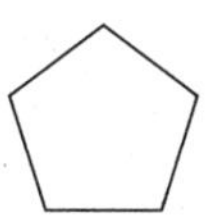

Rotational symmetry is when you can turn an object so that it looks exactly the same. The number of positions in which it looks exactly the same gives you its order of symmetry.

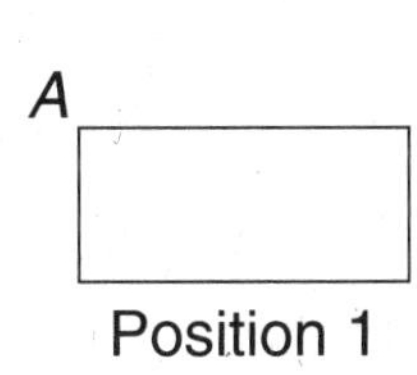

Position 1

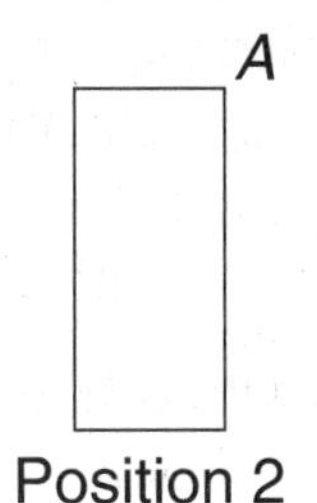

Position 2

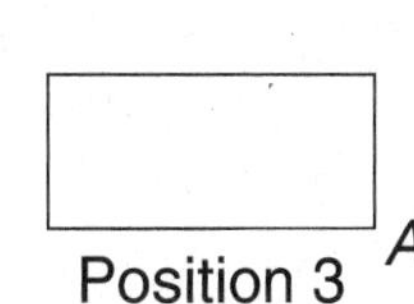

Position 3

Follow vertex *A* to see how the figure is turned.
Just twice, the figure looks exactly the same, in Positions 1 and 3.
So, this figure has 2-fold rotational symmetry.

Tell the order of rotational symmetry.

4. 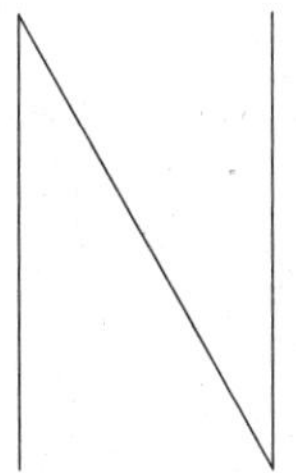

____

5. 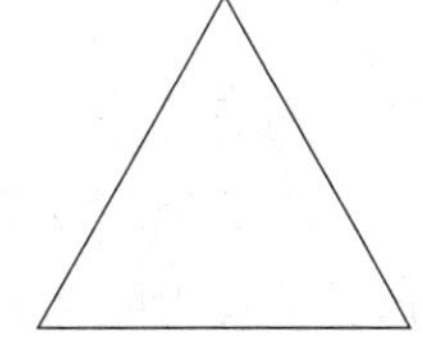

____

6. 

____

Name ______________________ Date ____________ Class ____________

LESSON 7-8

Challenge
Inside, Outside

Materials: sheet of paper (8.5 in. by 11 in.), scissors, tape

1. a. Cut a strip of paper 2 in. by 11 in. Tape the ends together to form a loop.

b. Use a pencil to create a center line around the outside of the loop. Cut the loop along its center line. How many loops? ______

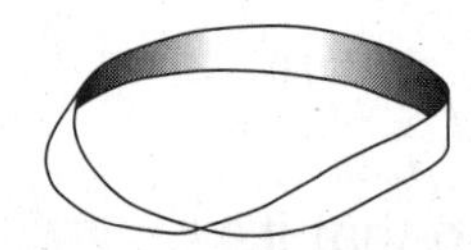

2. a. Cut a strip of paper 2 in. by 11 in. Before taping the ends together, give the strip a half-twist by turning one end over 180°.

b. Use a pencil to create a center line around the loop. Continue drawing until you reach your starting point. Describe the result. How many surfaces (sides) does this strip have?

____

____

c. Cut the loop along its center line. How many loops do you have?

____

This unusual surface is called a **Mobius Strip,** named after A. F. Mobius, a 19th century German mathematician and astronomer, who pioneered *topology* (how geometric figures act when distorted by such things as twists and stretches).

3. Prepare another Mobius Strip, a loop with a half-twist. Cut this strip parallel to one edge about one-third of the way from the edge. Continue cutting until you get back to your starting point. Describe the result.

____

4. Prepare a loop with a full twist. Cut this strip parallel to one edge about one-third of the way from the edge. Continue cutting until you get back to your starting point. Describe the result.

____

____

5. Prepare a loop with three half-twists. Cut this strip parallel to one edge about one-half of the way from the edge. Continue cutting until you get back to your starting point. Describe the result.

____

Name ______________________ Date __________ Class __________

LESSON 7-8

Problem Solving
Symmetry

Complete the figure. A dashed line is a line of symmetry and a point is a center of rotation.

1.

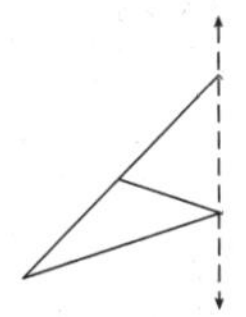

2.

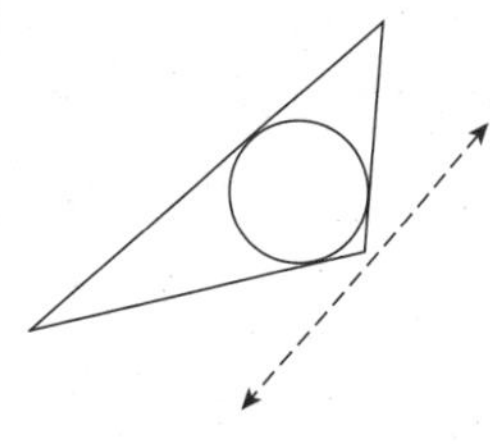

3. 2 fold symmetry

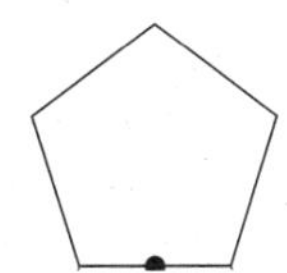

4. 4 fold symmetry

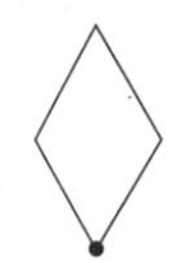

Use the flag of Switzerland to answer the questions.

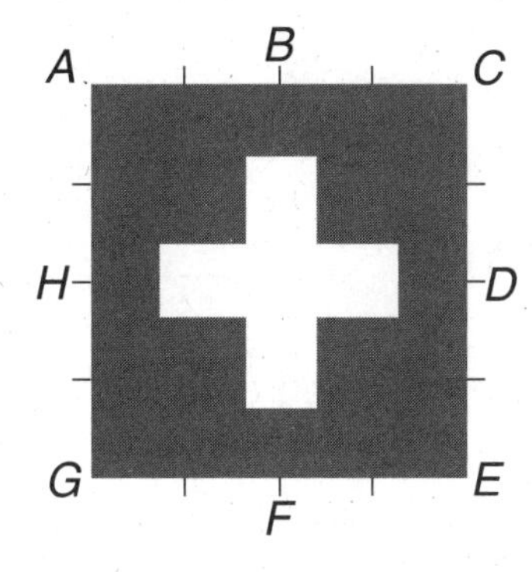

5. Which of the following would NOT be a line of symmetry?

A $\overline{HD}$ **C** $\overline{AE}$

B $\overline{BF}$ **D** $\overline{HB}$

6. How many lines of symmetry does the flag have?

F 2 **H** 4

G 6 **J** 8

7. How many folds of rotational symmetry does the flag have?

A 0 **C** 2

B 4 **D** 8

8. Which lists all lines of symmetry of the flag?

F $\overline{AE}$, $\overline{GC}$

G $\overline{HD}$, $\overline{BF}$

H $\overline{HD}$, $\overline{BF}$, $\overline{AE}$, $\overline{GC}$

J $\overline{HB}$, $\overline{DF}$, $\overline{AE}$, $\overline{GC}$

9. Which describes the center of rotation?

A intersection of $\overline{BF}$ and $\overline{HD}$

B intersection of $\overline{AE}$ and $\overline{HB}$

C *A*

D There is no center of rotation

Name ______________________ Date ___________ Class ___________

LESSON 7-8

Math: Reading and Writing in the Content Area

Reading a Diagram

The star below turns around a center point. If a figure rotates onto itself before one complete rotation, the figure has **rotational symmetry.**

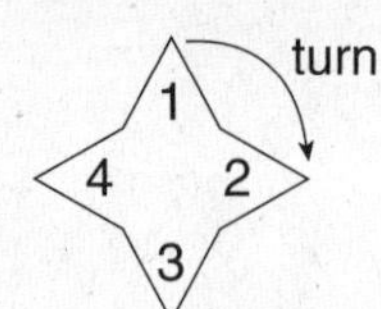

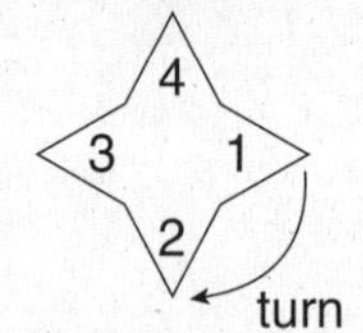

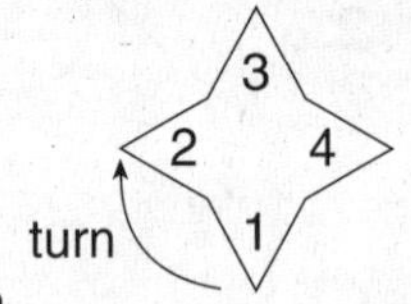

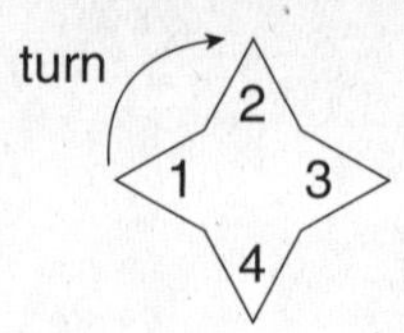

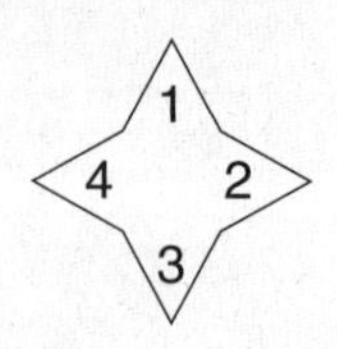

The star matches up with the original figure before it has made one complete turn. It has rotational symmetry.

This right triangle does not have rotational symmetry. The rotated image matches up with the original triangle only after it has made one complete turn.

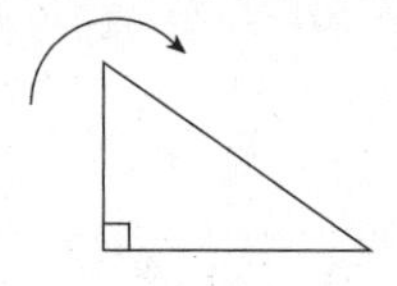

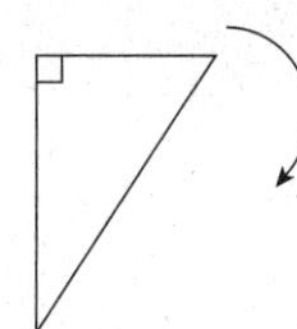

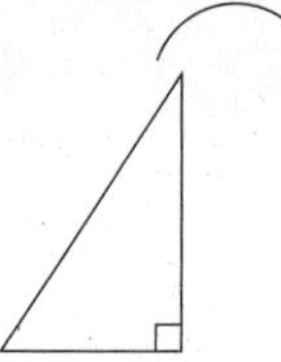

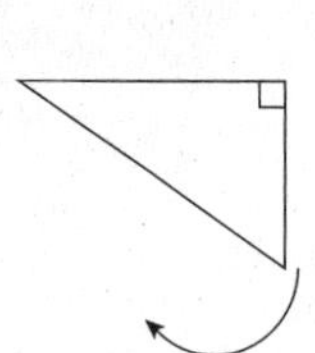

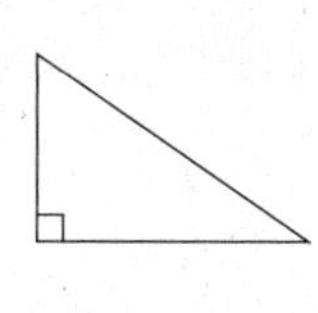

Do these figures have rotational symmetry? Trace each figure and rotate it to find out. Write *yes* or *no.*

1.

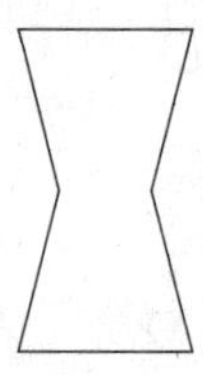

____

2.

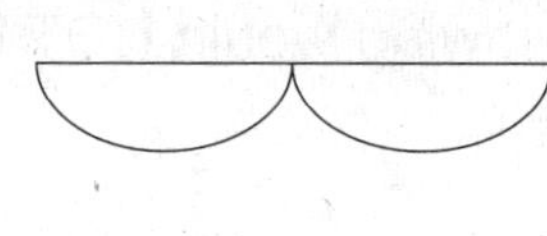

____

A figure can be reflected over a line so that the two parts are congruent. That figure has **line symmetry.**

Line Symmetry

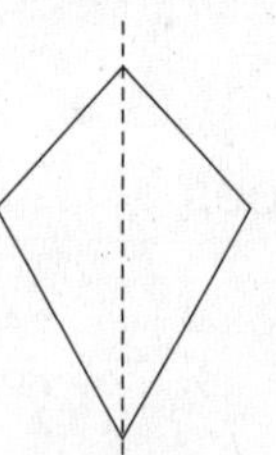

No Line Symmetry

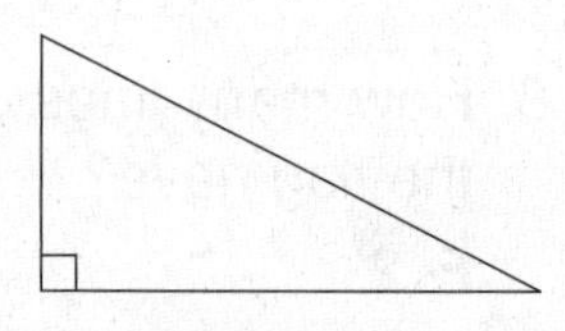

Do these figures have line symmetry? Trace each figure. Draw lines of symmetry if possible. Write *yes* or *no.*

3.

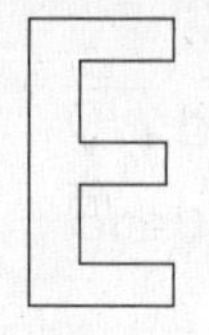

____

4.

____

Name ________________________ Date __________ Class __________

LESSON **7-8**

Puzzles, Twisters & Teasers

Concentrating on Symmetrical Figures

Imagine a game of concentration. Each box represents a card in the game. Choose two cards with figures that would have symmetry if they were put together. Cross them out. Rearrange the bold letters of the unmatched cards to solve the riddle.

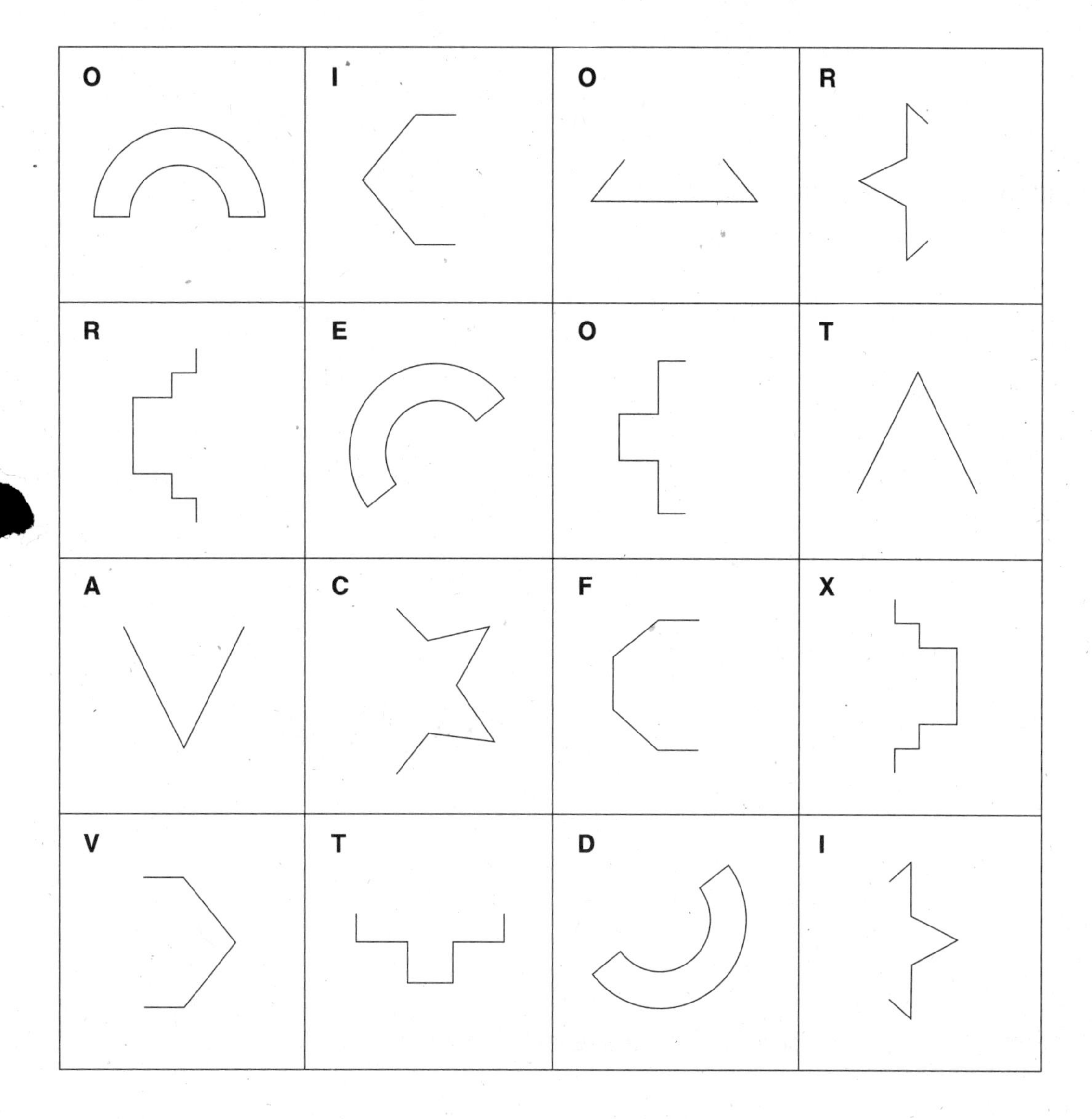

Who earns a living by driving his customers away?

A ___ ___ ___ ___   ___ ___ ___ ___ ___ ___

Name ______________________ Date ___________ Class ___________

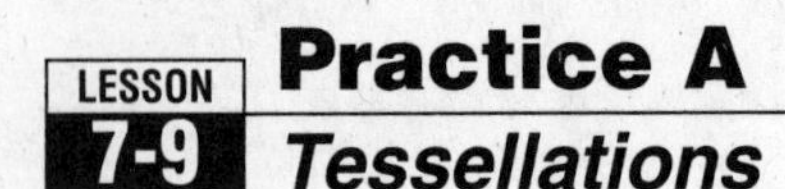

Practice A
LESSON 7-9 Tessellations

1. Create a tessellation with triangle JKL.

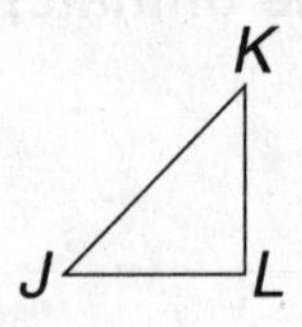

2. Use rotations to create a variation of the tessellation in Exercise 1.

3. Create a tessellation with quadrilateral *ABCD*.

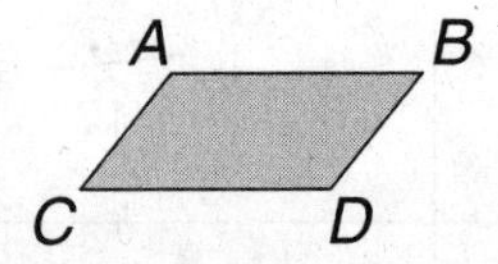

4. Use rotations to create a variation of the tessellation in Exercise 3.

Name ______________________ Date __________ Class __________

Practice B
Tessellations

1. Create a tessellation with quadrilateral *ABCD*.

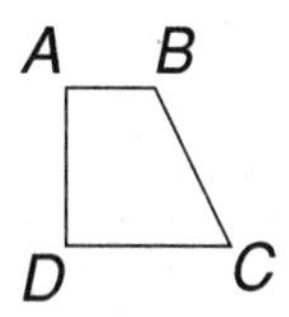

2. Use rotations to create a variation of the tessellation in Exercise 1.

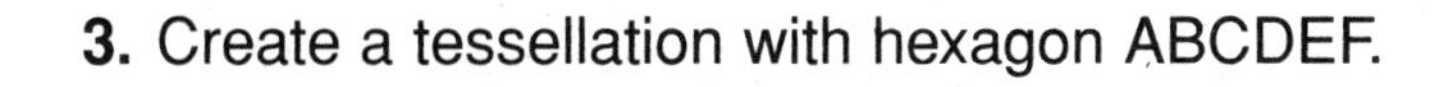

3. Create a tessellation with hexagon ABCDEF.

A B
F C
E D

4. Use rotations to create a variation of the tessellation in Exercise 3.

Name ______________________ Date ____________ Class ____________

LESSON 7-9

Practice C

Tessellations

1. Create a tessellation with triangle *ABC*.

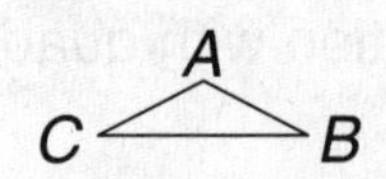

2. Use rotation to create a variation of the tessellation in Exercise 1.

3. Use a rhombus and a parallelogram to create a tessellation.

Copy each shape and use it to create a tessellation.

4.

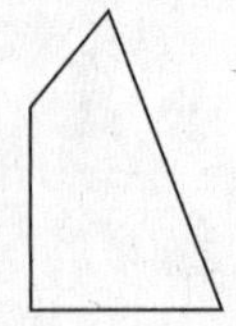

5.

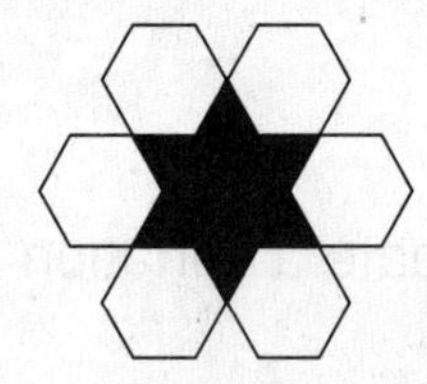

Name ______________________________ Date __________ Class __________

Reteach

Tessellations

A **tessellation** is a repeating pattern of shapes that completely covers a plane with no gaps or overlaps.

Not all plane shapes can be used to make a tessellation.

To make a tessellation of a figure, fit copies of the shape together.
Remember, there can be no overlaps or gaps when you fit the shapes together.

You can make a tessellation using this hexagon.

The figures fit together with no gaps or overlaps.

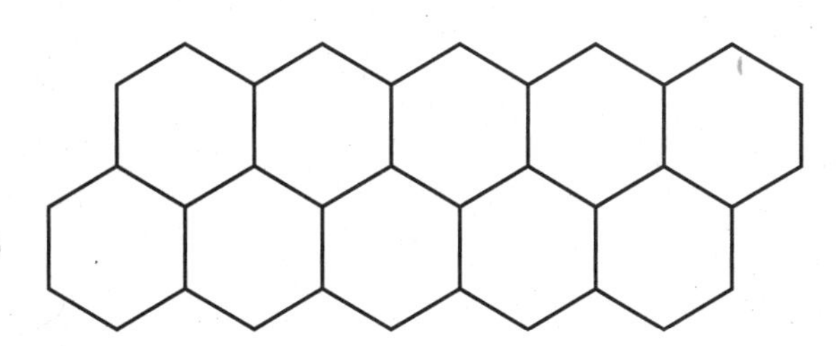

You cannot make a tessellation using this pentagon.

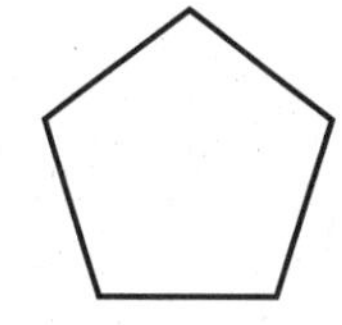

The figures do not all fit together. There is a gap.

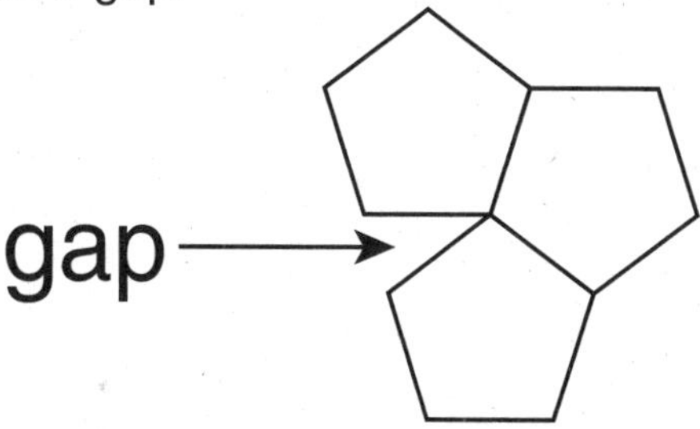

Can you make a tessellation using each shape? Write yes or no.

1.

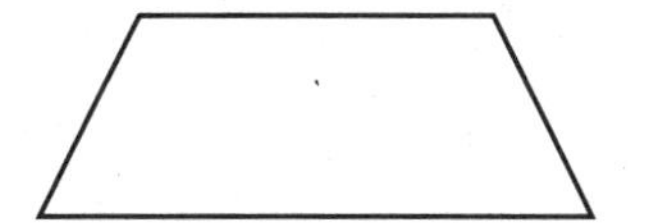

____

2.

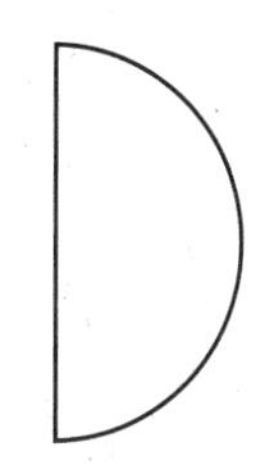

____

Fit triangles shaped like this together to make a tessellation.

3.

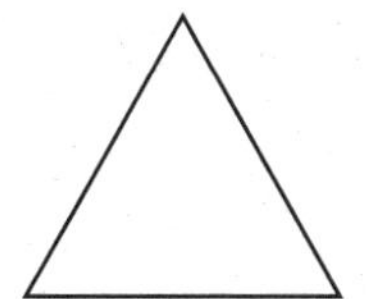

Name ______________________ Date ____________ Class ____________

LESSON 7-9

Reteach

Tessellations (continued)

After making a **tessellation** of a figure, you can use a rotation to create a variation of the tessellation.

Use the figure shown at the right to create a tessellation.

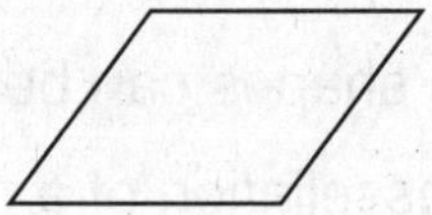

Tessellation

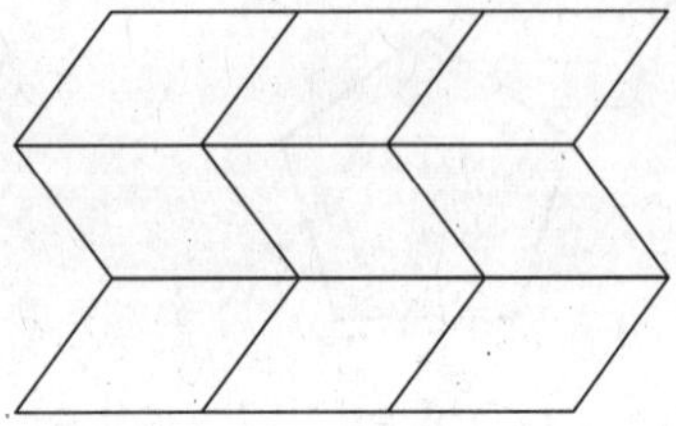

Variation

Rotate the tessellation 90° clockwise to vary the tessellation.

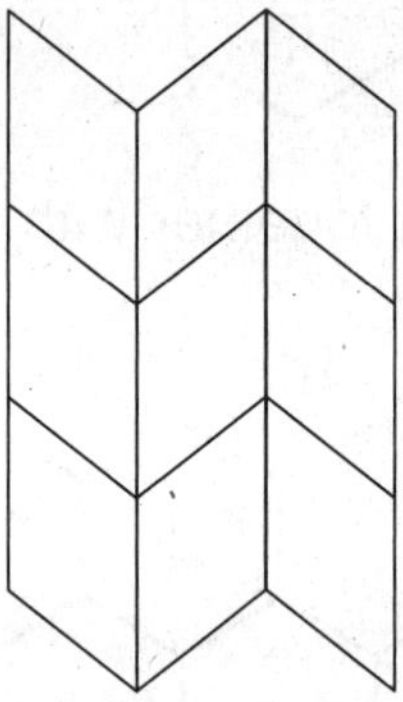

4. Make a tessellation using the shape?

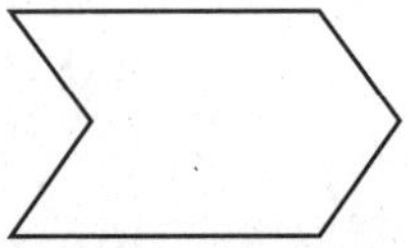

5. Rotate your tessellation 90° clockwise to vary the tessellation.

Name ______________________ Date __________ Class __________

LESSON 7-9 Challenge

Shapely

A **fractal** is a shape that repeats itself in a pattern. As more stages are generated, the shape becomes more complex.

Starting with a + sign and generating the pattern by adding a half-size + sign in each of the four corners, the first two stages of a fractal are shown.

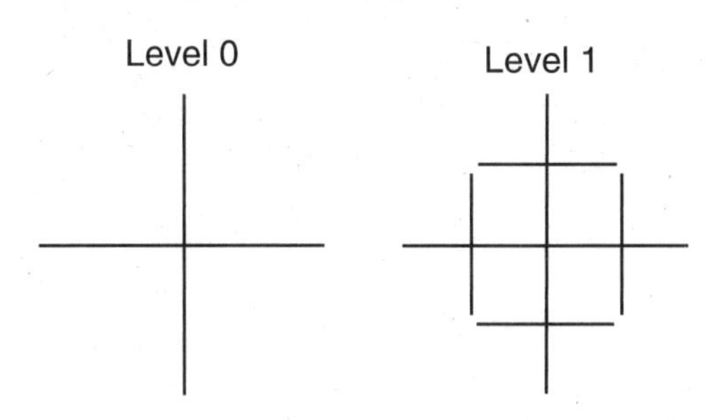

1. Generate the next two stages of the fractal.

   Level 2 Level 3

Use paper folding to generate a fractal.
Level 0: Begin with a long strip of paper.

2. Level 1: Fold the strip right end to left; crease it. Open the strip so that it forms a right angle. Stand the strip on edge on your desk. Look down and sketch this view.

3. Level 2: Start at Level 0. Do Level 1 (the pattern) twice. Open the strip so that it forms right angles. Stand the strip on edge on your desk. Look down and sketch this view. Comment on the result.

   ____

   ____

4. Level 3: Start at Level 0. Do the pattern three times.

5. Level 4: Start at Level 0. Do the pattern four times.

Name ______________________ Date ____________ Class ____________

LESSON 7-9

Problem Solving

Tessellations

Create a tessellation using the given figure.

1. 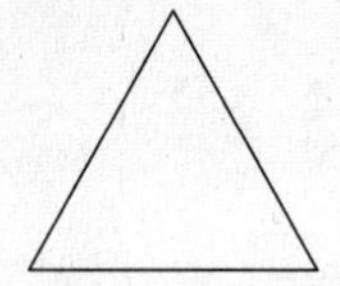

2. 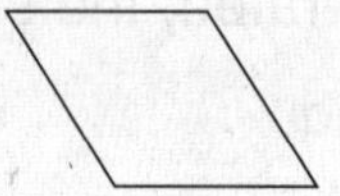

Choose the letter for the best answer.

3. Which figure will NOT make a tessellation?

A

B

C

D

4. Which nonregular polygon can always be used to tile a floor?

F pentagon

G triangle

H octagon

J hexagon

5. For a combination of regular polygons to tessellate, the angles that meet at each vertex must add to what?

A 90°

B 180°

C 360°

D 720°

Name ______________________________ Date __________ Class ___________

Reading Strategies

Draw Conclusions

When figures are arranged in a repeating pattern that covers a surface with no gaps or overlaps, the pattern is called a **tessellation.** This tessellation is made with equilateral triangles.

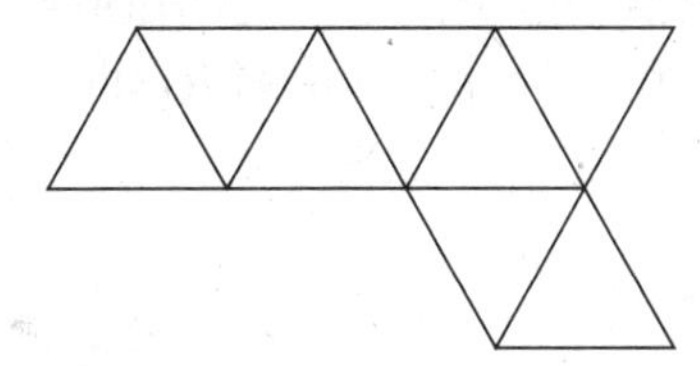

Repeating regular pentagons do not cover a surface completely. There are gaps between the pentagons, so this figure does not tessellate.

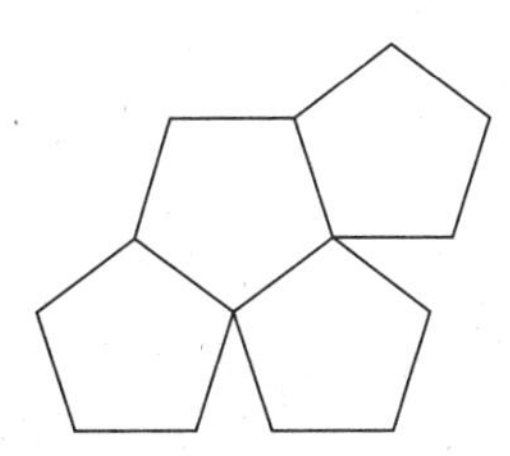

Trace and cut out several of the figures below. Test to see if the figures fit together to form a tessellation. Write *yes* or *no.*

1. 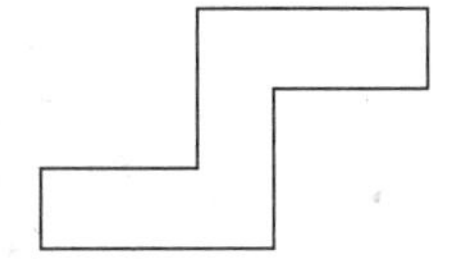

____

2. 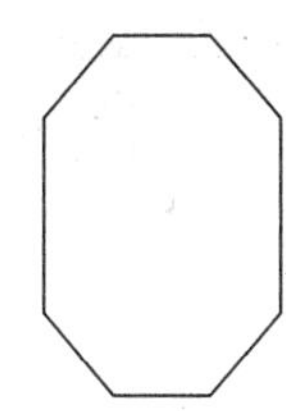

____

3. 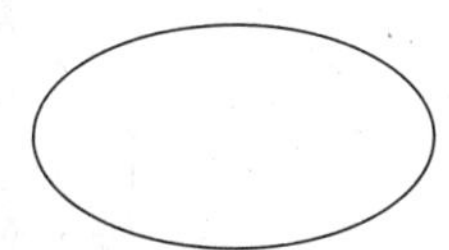

____

4. 

____

Name ______________________ Date __________ Class __________

LESSON 7-9

Puzzles, Twisters & Teasers

A Puzzling Pattern!

Solve the crossword puzzle.

Across

1. A _____ is a repeating pattern of a plane figure that covers a plane with no gaps or overlaps.

5. A quadrilateral with 4 equal sides and 4 equal angles is a _____.

6. To move a figure around a central point is to _____.

7. A 6-sided polygon is a _____.

8. The angle measures of a _____ add to 360°.

9. A repeating _____ is a design created by using the same figure in sequence over and over.

Down

2. In a _____ tessellation, two or more regular polygons are repeated to fill the plane.

3. An 8-sided polygon is an _____.

4. When the same figure is copied over and over again, it is _____.

6. In a _____ polygon, all sides are equal.

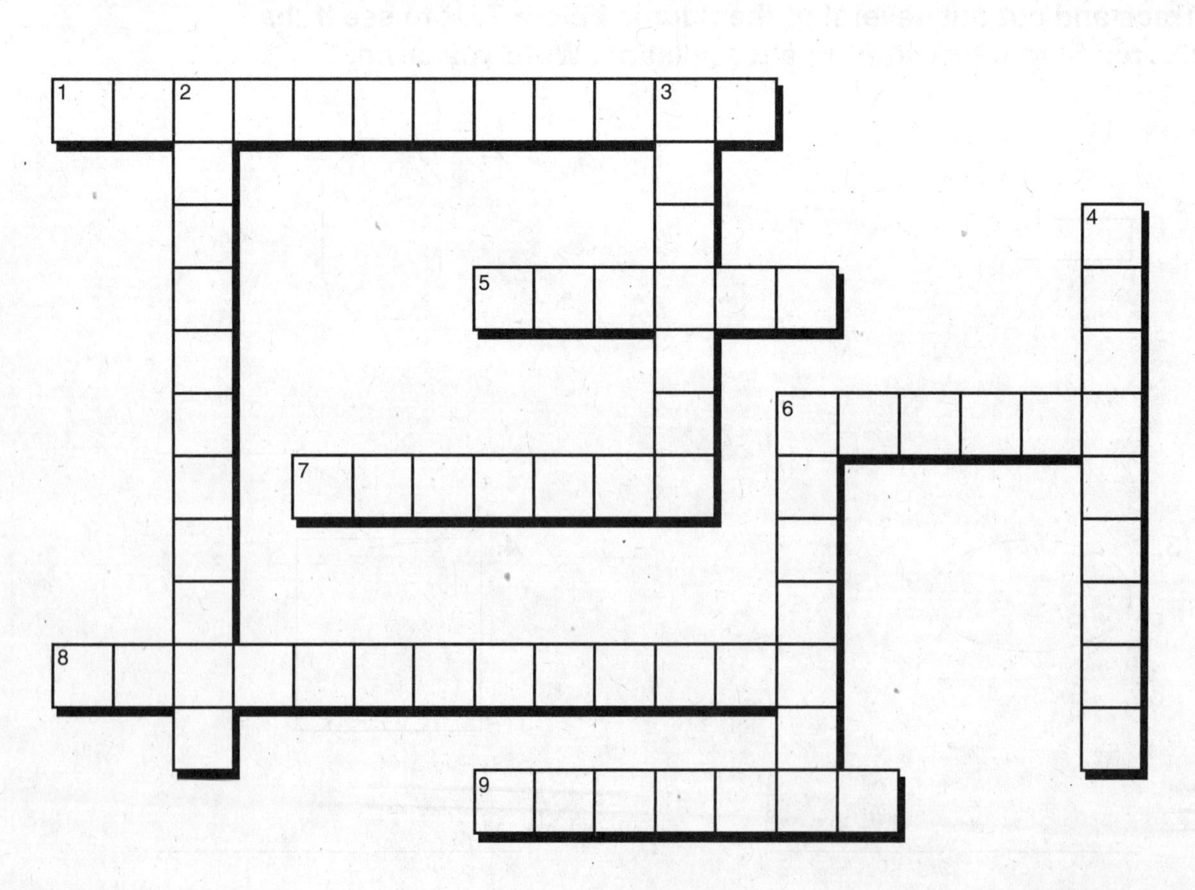

LESSON 7-1

Practice A

Points, Lines, Planes, and Angles

Use the diagram to name each figure.

1. four points

   A, B, C, D

2. a line

   $\overleftrightarrow{AD}$

3. a plane

   Possible answer: *ABC*

4. three segments

   $\overline{AB}$, $\overline{BC}$, $\overline{CD}$, or $\overline{AD}$

5. four rays

   $\overrightarrow{AB}$, $\overrightarrow{AD}$, $\overrightarrow{DA}$, $\overrightarrow{DC}$

Use the diagram to name each figure.

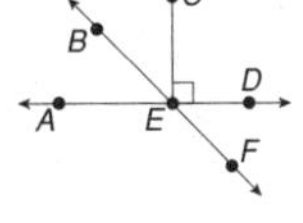

6. a right angle

   $\angle CED$* or *$\angle AEC$

7. two acute angles

   $\angle AEB$, $\angle BEC$*, or *$\angle DEF$

8. two obtuse angles

   Possible answer: *$\angle BED$* and *$\angle CEF$*

9. pair of complementary angles

   $\angle AEB$* and *$\angle BEC$

10. two pairs of supplementary angles

    Possible answer: *$\angle AEC$* and *$\angle CED$*, *$\angle BEC$* and *$\angle CEF$*

In the figure, $\angle 1$ and $\angle 3$ are vertical angles, and $\angle 2$ and $\angle 4$ are vertical angles.

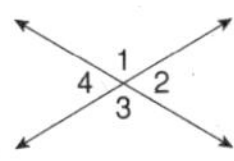

11. If $m\angle 1 = 120°$, find $m\angle 3$.

    $m\angle 3 = 120°$

12. If $m\angle 2 = x°$, find $m\angle 4$.

    $m\angle 4 = x°$

LESSON 7-1

Practice B

Points, Lines, Planes, and Angles

Use the diagram to name each figure.

1. four points

   R, S, T, W

2. a line

   $\overleftrightarrow{WT}$

3. a plane

   Possible answer: *RST*

4. three segments

   $\overline{RS}$, $\overline{ST}$, $\overline{TW}$, or $\overline{RW}$

5. four rays

   $\overrightarrow{RS}$, $\overrightarrow{WT}$, $\overrightarrow{WR}$, $\overrightarrow{TW}$

Use the diagram to name each figure.

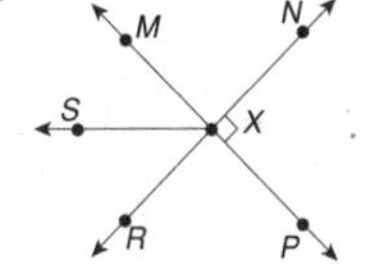

6. a right angle

   $\angle NXP$, $\angle MXR$, $\angle MXN$*, or *$\angle RXP$

7. two acute angles

   $\angle MXS$* and *$\angle RXS$

8. two obtuse angles

   $\angle NXS$* and *$\angle PXS$

9. a pair of complementary angles

   $\angle RXS$* and *$\angle SXM$

10. three pairs of supplementary angles

    Possible answer: *$\angle NXP$* and *$\angle MXN$*, *$\angle RXM$* and *$\angle MXN$*, *$\angle RXP$* and *$\angle PXN$*

In the figure, $\angle 1$ and $\angle 3$ are vertical angles, and $\angle 2$ and $\angle 4$ are vertical angles.

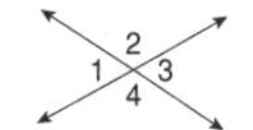

11. If $m\angle 2 = 110°$, find $m\angle 4$.

    $m\angle 4 = 110°$

12. If $m\angle 1 = n°$, find $m\angle 3$.

    $m\angle 3 = n°$

LESSON 7-1

Practice C

Points, Lines, Planes, and Angles

Use the diagram to solve.

Write *true* or *false*. If a statement is false, rewrite it so it is true.

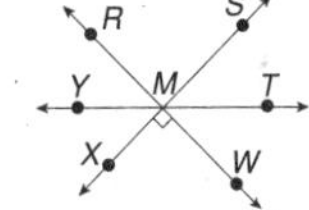

1. *RW* is a line.

   true

2. Rays *YM* and *MS* make up line *YT*.

   false; Rays *YM* and *TM* make up line *YT*

3. Angle *RMT* is an obtuse angle.

   true

4. $\angle WMY$ and $\angle RMY$ are supplementary angles.

   true

5. $\angle YMX$ and $\angle SMT$ are supplementary angles.

   false; *$\angle YMX$* and *$\angle SMT$* are vertical angles

6. $\angle XMY$ and $\angle YMR$ are complementary angles.

   true

7. If $m\angle SMT = 48°$, then $m\angle TMW = 48°$.

   false; If $m\angle SMT = 48°$, then $m\angle TMW = 42°$

8. If $m\angle SMY = 135°$, then $m\angle RMT = 135°$.

   true

9. If $m\angle TMW = x°$, then $m\angle RMT = 180° - x°$.

   true

10. $m\angle RMY + m\angle SMT + m\angle RMS = 180°$.

    true

In the figure, $\angle 1$ and $\angle 3$ are vertical angles.

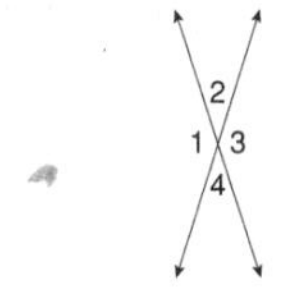

11. If $m\angle 1 = 123°$, find $m\angle 3$.

    123°

12. If $m\angle 1 + m\angle 2 = 180°$ and $m\angle 2 = 40°$, find $m\angle 3$.

    140°

LESSON 7-1

Reteach

Points, Lines, Planes, and Angles

Figure	Description	Diagram	Notation Write	Notation Read
Line	an infinite collection of points with no beginning and no end	ℓ, A, B	$\overleftrightarrow{AB}$ or $\overleftrightarrow{BA}$ or ℓ	line *AB*, line *BA*, line ℓ
Line Segment	part of a line, with two endpoints	A, B	$\overline{AB}$ or $\overline{BA}$	line segement *AB*, line segment *BA*
Ray	part of a line, with one endpoint	A, B	$\overrightarrow{AB}$	ray *AB*

Use the diagram, to name each type of figure.

1. $\overrightarrow{MP}$ **ray**
2. *k* **line**
3. $\overline{MN}$ **line segment**
4. $\overline{LJ}$ **line segment**
5. $\overleftrightarrow{JL}$ **line**

Acute Angle	Right Angle	Obtuse Angle	Straight Angle
Measures between 0° and 90°	Measures exactly 90°	Measures between 90° and 180°	Measures exactly 180°.

Use the diagram to name each type of angle.

6. $\angle BCD$ **right angle**
7. $\angle BAD$ **acute angle**
8. $\angle BDA$ **obtuse angle**
9. $\angle CDA$ **straight angle**
10. $\angle BDC$ **acute angle**
11. $\angle ABC$ **acute angle**

LESSON 7-1 **Reteach**

Points, Lines, Planes, and Angles (continued)

Complementary Angles	Supplementary Angles	Vertical Angles

Two angles whose measures have a sum of 90°.	Two angles whose measures have a sum of 180°.	Intersecting lines form two pairs of vertical angles.
$\angle A$ and $\angle B$ are complementary angles.	$\angle C$ and $\angle D$ are supplementary angles.	$\angle a$ and $\angle b$, $\angle c$ and $\angle d$ are pairs of vertical angles.

Use the diagram to complete.

12. Since $\angle AQC$ and $\angle DQB$ are formed by intersecting lines, $\overleftrightarrow{AQB}$ and $\overleftrightarrow{CQD}$, they are:

vertical angles

13. The sum of the measures of $\angle AQV$ and $\angle VQT$ is: 90°

So, these angles are:

complementary angles

14. The sum of the measures of $\angle AQC$ and $\angle CQB$ is: 180°

So, these angles are: supplementary angles

Congruent figures have the same size and shape.
The symbol ≅ means *is congruent to.*

Complete these statements about congruence.

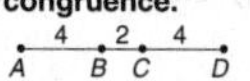

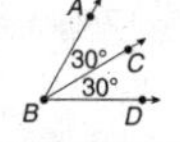

15. Since $AC = 6$ units and $BD = 6$ units, then $\overline{AC}$ ≅ $\overline{BD}$.

16. Since $m\angle ABC = 30°$ and $m\angle CBD = 30°$ then $\angle ABC$ ≅ $\angle CBD$.

17. Since vertical angles are congruent, then $\angle UPJ$ ≅ $\angle VPS$.

LESSON 7-1 **Challenge**

Let's Meet!

Materials needed: paper strips, index cards, and scissors

1. Use a flat surface such as the top of your desk to represent a plane. Use strips of paper to represent lines. Move the lines around in the plane (**coplanar lines**) to determine the number of intersections that are possible. Summarize your results in a table.

Number of Coplanar Lines	Possible Number of Points of Intersection
2	0 or 1
3	0, 1, 2, or 3
4	0, 1, 3, 4, 5, or 6
5	0, 1, 4, 5, 6, 7, 8, 9, or 10

2. Slit one index card and connect two cards to model two intersecting planes.

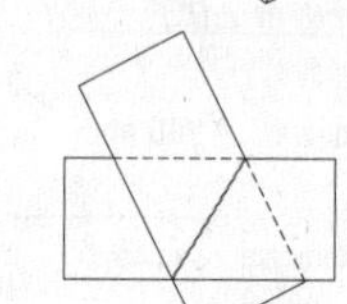

a. What is the intersection of two planes?

a line

b. Mark the diagram to illustrate the intersection of the two planes.

3. Using index cards to represent planes, determine the number of intersections that are possible. Summarize your results in a table.

Number of Planes	Possible Number of Lines of Intersection
2	0 or 1
3	0, 1, 2, or 3
4	0, 1, 3, 4, 5, or 6
5	0, 1, 4, 5, 6, 7, 8, 9, or 10

LESSON 7-1 **Problem Solving**

Points, Lines, Planes, and Angles

Use the flag of the Bahamas to solve the problems.

1. Name four points in the flag.

Possible answers: A, B, C, D

2. Name four segments in the flag.

Possible answers: $\overline{AB}$, $\overline{BH}$, $\overline{HI}$, $\overline{IC}$

3. Name a right angle in the flag.

Possible answer: $\angle DAB$

4. Name two acute angles in the flag.

Possible answers: $\angle AED$, $\angle DAE$

5. Name a pair of complementary angles in the flag.

Possible answer: $\angle DAE$, $\angle EAB$

6. Name a pair of supplementary angles in the flag.

Possible answer: $\angle DGI$, $\angle IGE$

The diagram illustrates a ray of light being reflected off a mirror. The angle of incidence is congruent to the angle of reflection. Choose the letter for the best answer.

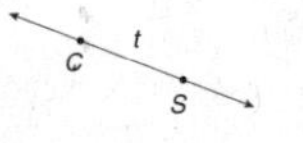

7. Name two rays in the diagram.

A $\overrightarrow{AM}$, $\overrightarrow{MB}$ (C) $\overrightarrow{MA}$, $\overrightarrow{MB}$
B $\overrightarrow{MA}$, $\overrightarrow{BM}$ D $\overline{MA}$, $\overline{MB}$

8. Name a pair of complementary angles.

(F) $\angle NMB$, $\angle BMD$ H $\angle CMA$, $\angle AMD$
G $\angle AMN$, $\angle NMB$ J $\angle CMA$, $\angle DMB$

9. Which angle is congruent to $\angle 2$?

A $\angle 1$ (C) $\angle 3$
B $\angle 4$ D none

10. Find the measure of $\angle 4$.

F 65°   (H) 25°
G 35°   J 90°

11. Find the measure of $\angle 1$.

A 65°   (C) 25°
B 35°   D 90°

12. Find the measure of $\angle 3$.

F 90°   H 35°
G 45°   (J) 65°

LESSON 7-1 **Reading Strategies**

Understanding Vocabulary

A **line** is a straight path that extends forever in both directions. A line can be named in two different ways.

- A lowercase letter can name a line.
- Two points on a line can name the line.

A **line segment** is part of a line running between two endpoints. A line segment is named by its endpoints, using capital letters.

Line t or $\overleftrightarrow{CS}$ or $\overleftrightarrow{SC}$
Read: line t, or line CS, or line SC

$\overline{CS}$ or $\overline{SC}$
Read: segment CS or segment SC

Use this figure to answer each question.

1. How would you name the line segment?

segment EF or segment FE

2. How would you name the line in this figure?

line r or line EF or line FE

A **plane** is a flat surface that extends without end in all directions. You use three points that are not on a line to name a plane. Planes are named using capital letters. This plane could be named plane *MVW*.

3. What is another way to name this plane?

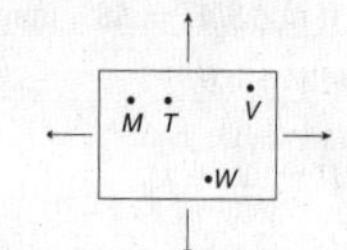

Possible answer: TVW

4. Draw a plane and identify three points in the plane that are not on a line.

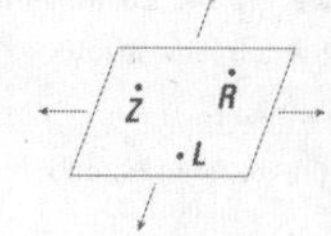

Possible answer: plane ZRL; Z, R, and L

LESSON 7-1

Puzzles, Twisters & Teasers

What's Your Angle?

Circle the words below in the word search (horizontally, vertically or diagonally). Then find a word in the word search that answers the riddle. Circle it and write it on the line.

point	supplementary	plane	segment	ray
angle	complementary	line	congruent	vertical

```
C O M P L E M E N T A R Y S
O M H O Y U P U I O N W E U
N O L I T Y L T R Y G R T P
G I U N R T A W E R L U I P
R A Y T Q W N P L M E Y H L
U W D V G T E Q W E R T Y E
E E D C V F R T G B R H Y M
N O I U Y T R F V B A J K E
T I S E G M E N T U Y T F N
I M L P O K N B J L U H V T
M W D C F R T G H N I J U A
Q V E R T I C A L D C N H R
E P A N T S U J M K I O E Y
```

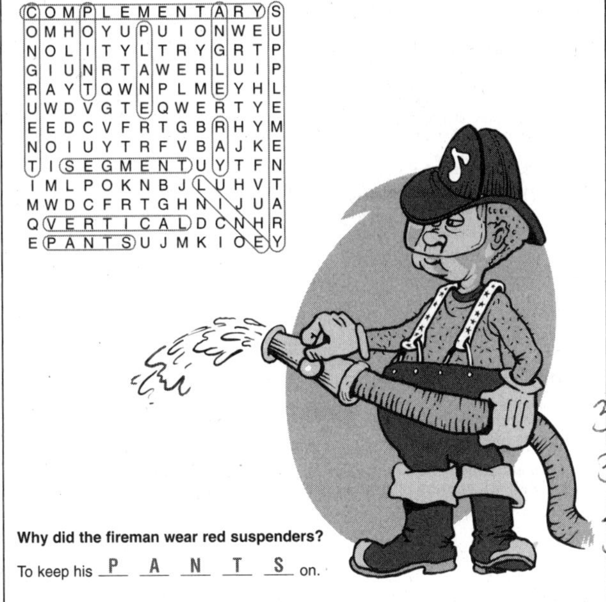

Why did the fireman wear red suspenders?

To keep his P A N T S on.

LESSON 7-2

Practice A

Parallel and Perpendicular Lines

1. Measure the angles formed by the transversal and the parallel lines. Which angles seem to be congruent?
   ∠1 ≅ ∠3 ≅ ∠5 ≅ ∠7 and ∠2 ≅ ∠4 ≅ ∠6 ≅ ∠8

In the figure, line *r* ∥ line *s*. Find the measure of each angle.

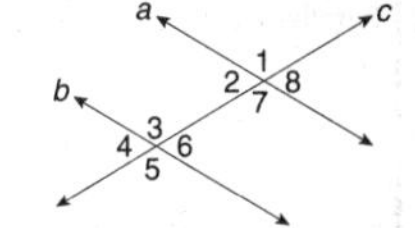

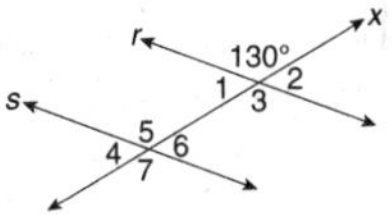

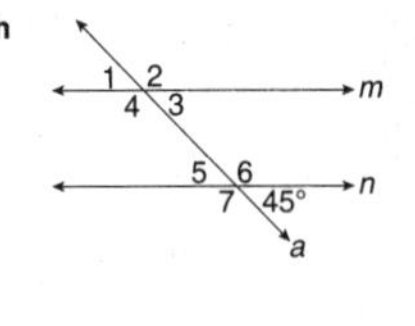

2. ∠1 50°
3. ∠5 130°
4. ∠6 50°
5. ∠7 130°
6. ∠4 50°
7. ∠3 130°

In the figure, line *m* ∥ line *n*. Find the measure of each angle.

8. ∠1 45°
9. ∠2 135°
10. ∠3 45°
11. ∠5 45°
12. ∠6 135°
13. ∠7 135°

In the figure, line *a* ∥ line *b*.

14. Name all angles congruent to ∠1.
    ∠3, ∠5, ∠7
15. Name all angles congruent to ∠2.
    ∠4, ∠6, ∠8
16. Name three pairs of supplementary angles.
    Possible answer: ∠1 + ∠8, ∠2 + ∠7, ∠3 + ∠6, or ∠4 + ∠5
17. Which line is the transversal?
    line *c*

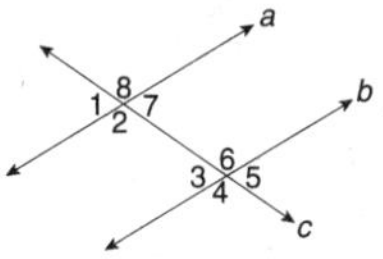

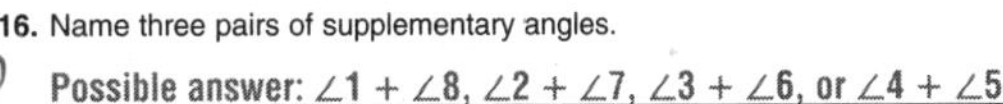

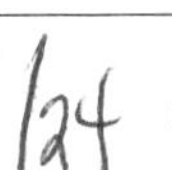

LESSON 7-2

Practice B

Parallel and Perpendicular Lines

1. Measure the angles formed by the transversal and the parallel lines. Which angles seem to be congruent?
   ∠1 ≅ ∠4 ≅ ∠5 ≅ ∠8 and ∠2 ≅ ∠3 ≅ ∠6 ≅ ∠7

In the figure, line *m* ∥ line *n*. Find the measure of each angle.

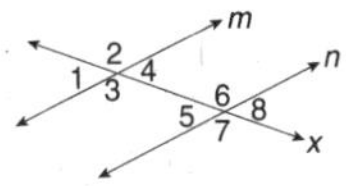

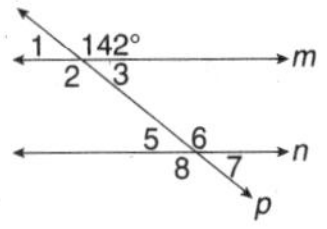

2. ∠1 38°
3. ∠2 142°
4. ∠5 38°
5. ∠6 142°
6. ∠8 142°
7. ∠7 38°

In the figure, line *a* ∥ line *b*. Find the measure of each angle.

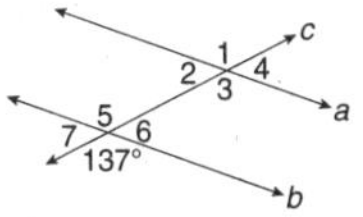

8. ∠2 43°
9. ∠5 137°
10. ∠6 43°
11. ∠7 43°
12. ∠4 43°
13. ∠3 137°

In the figure, line *r* ∥ line *s*.

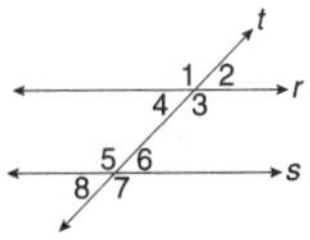

14. Name all angles congruent to ∠2.
    ∠4, ∠6, ∠8
15. Name all angles congruent to ∠7.
    ∠1, ∠3, ∠5
16. Name three pairs of supplementary angles.
    Possible answer: ∠1 + ∠2, ∠3 + ∠4, ∠5 + ∠6, or ∠7 + ∠8
17. Which line is the transversal?
    line *t*

LESSON 7-2

Practice C

Parallel and Perpendicular Lines

In the figure, line *a* ∥ line *b*.

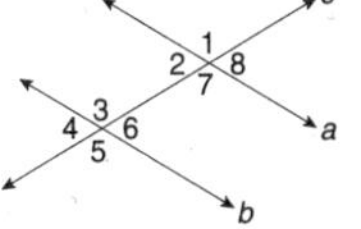

1. Name all angles congruent to ∠1.
   ∠3 ≅ ∠5 ≅ ∠7
2. Name all angles congruent to ∠2.
   ∠4 ≅ ∠6 ≅ ∠8
3. Name three pairs of supplementary angles.
   Possible answer: ∠1 + ∠8, ∠2 + ∠7, ∠3 + ∠6, or ∠4 + ∠5
4. Which line is the transversal?
   line *c*
5. If m∠7 is 131°, what is the m∠8?
   49°
6. If m∠4 is 57°, what is the m∠5?
   123°
7. If m∠3 is 127°, what is the m∠8?
   53°
8. If a transversal were drawn perpendicular to line *a* and line *b*, what would be the measure of the angles formed?
   90°

Draw a diagram to illustrate each of the following.

9. line *x* ∥ line *y* ∥ line *z* and transversal *k*
   sample answer shown

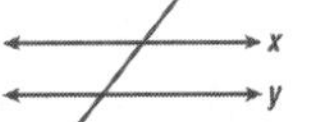

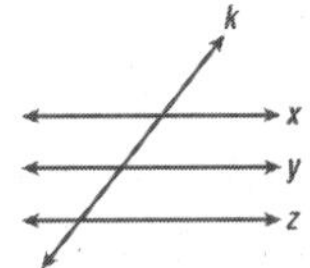

10. line *r* ∥ line *s* and transversal *m* with eight congruent angles
    sample answer shown

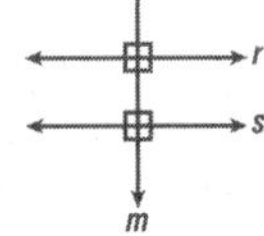

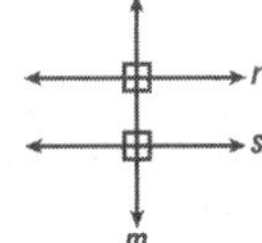

11. line *a* ∥ line *b* and transversal *t* with ∠1 ≅ ∠2 and ∠3 ≅ ∠4
    sample answer shown

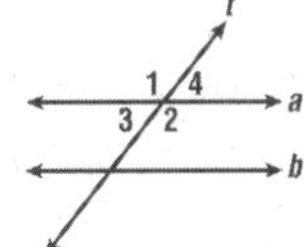

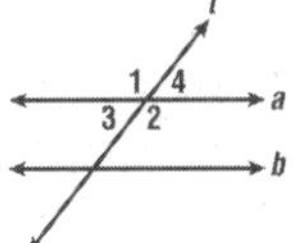

LESSON 7-2
Reteach
Parallel and Perpendicular Lines

Parallel Lines — Parallel lines never meet.

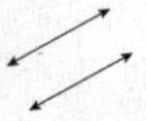

Perpendicular Lines — 90° — Perpendicular lines form right angles.

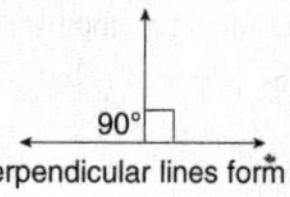

When parallel lines are cut by a **transversal,** 8 angles are formed, 4 acute and 4 obtuse.

The acute angles are all congruent.

The obtuse angles are all congruent.

Any acute angle is supplementary to any obtuse angle.

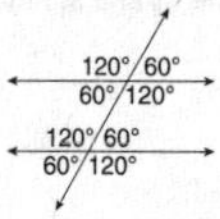

120° 60° / 60° 120° / 120° 60° / 60° 120°

In each diagram, parallel lines are cut by a transversal. Name the angles that are congruent to the indicated angle.

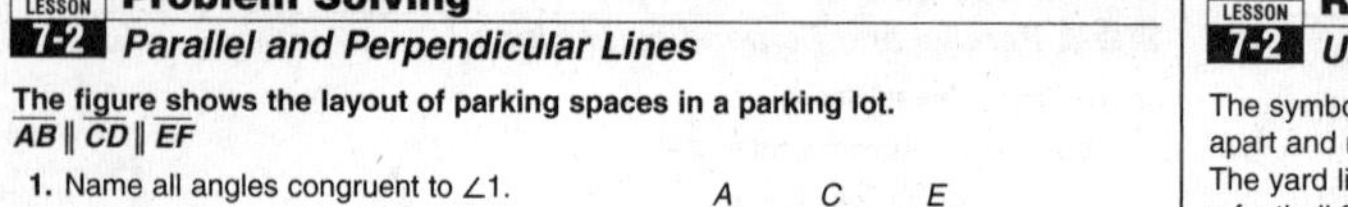

1. 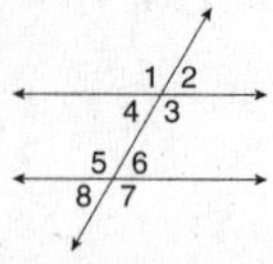 The angles congruent to ∠1 are: ∠3, ∠5, ∠7

2. 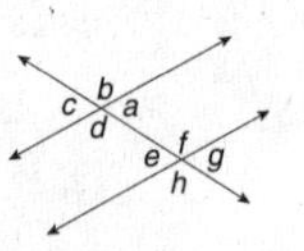 The angles congruent to ∠a are: ∠c, ∠e, ∠g

3. 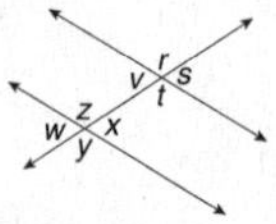 The angles congruent to ∠z are: ∠y, ∠t, ∠r

In each diagram, parallel lines are cut by a transversal and the measure of one angle is given. Write the measures of the remaining angles on the diagram.

4. 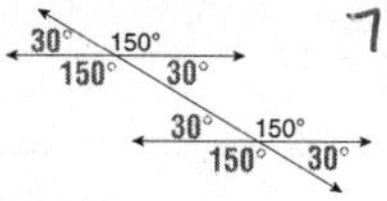 30° 150° / 150° 30° / 30° 150° / 150° 30°

5. 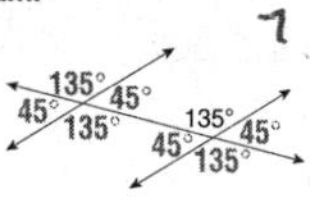 135° 45° / 45° 135° / 135° 45° / 45° 135°

6. 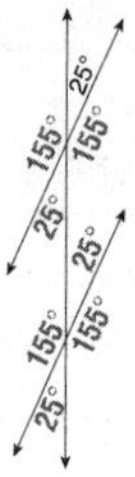 25° 155° / 155° 25° / 25° 155° / 155° 25°

LESSON 7-2
Challenge
Pairing Off

When two parallel lines are cut by a transversal, eight angles are formed. Of these, four angles are between the parallel lines, **interior angles.**

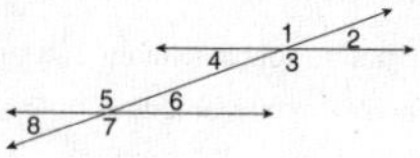

1. In this diagram, name the four interior angles formed by the parallel lines and the transversal.

∠3, ∠4, ∠5, ∠6

2. Think of the interior angles in pairs. Name the two pairs of interior angles that are on opposite sides of the transversal.

∠3 and ∠5; ∠4 and ∠6

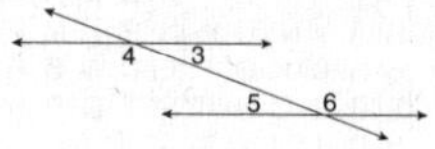

3. What is true about the measures of ∠3 and ∠5 in the diagram above? in the diagram at the right? Use a protractor to verify your conjecture.

∠3 ≅ ∠5

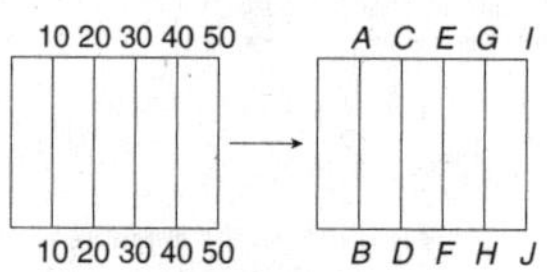

4. What is true about the measures of ∠4 and ∠6 in the diagram above? in the diagram at the right? Use a protractor to verify your conjecture.

∠4 ≅ ∠6

5. Interior angles that are on opposite sides of the transversal are called **alternate interior angles.**

Draw a conclusion about the measures of alternate interior angles formed by parallel lines and a transversal.

Alternate interior angles of parallel lines are equal in measure, or congruent.

Use your observation about the measures of alternate interior angles of parallel lines to find the measure of ∠x in each of these diagrams.

6. 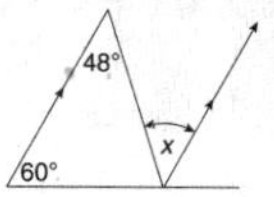 48°, 60°, x — m∠x = 48°

7. 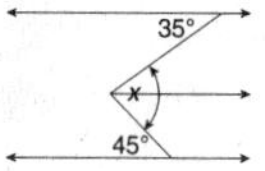 35°, x, 45° — m∠x = 35° + 45° = 80°

8. 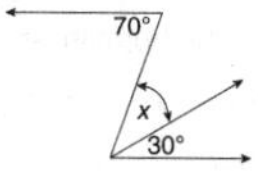 70°, x, 30° — m∠x = 70° − 30° = 40°

LESSON 7-2
Problem Solving
Parallel and Perpendicular Lines

The figure shows the layout of parking spaces in a parking lot. $\overline{AB} \parallel \overline{CD} \parallel \overline{EF}$

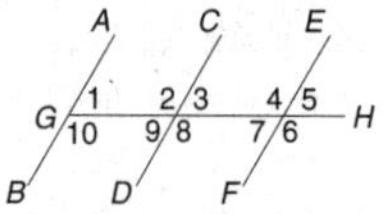

1. Name all angles congruent to ∠1.

∠3, ∠5, ∠7, ∠9

2. Name all angles congruent to ∠2.

∠4, ∠6, ∠8, ∠10

3. Name a pair of supplementary angles.

Possible answer: ∠1, ∠2

4. If m∠1 = 75°, find the measures of the other angles.

m∠3 = m∠5 = m∠7 = m∠9 = 75°, m∠2 = m∠4 = m∠6 = m∠8 = m∠10 = 105°

5. Name a pair of vertical angles.

Possible answer: ∠2, ∠8

6. If m∠1 = 90°, then $\overline{GH}$ is perpendicular to

Possible answers: $\overline{AB}$, $\overline{CD}$, $\overline{EF}$

The figure shows a board that will be cut along parallel segments *GB* and *CF*. $\overline{AD} \parallel \overline{HE}$. Choose the letter for the best answer.

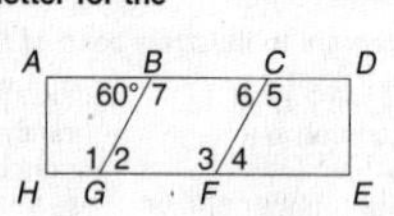

7. Find the measure of ∠1.
A 45° C 60°
(B) 120° D 90°

8. Find the measure of ∠2.
F 30° (H) 60°
G 120° J 90°

9. Find the measure of ∠3.
A 30° C 60°
(B) 120° D 90°

10. Find the measure of ∠4.
F 45° (H) 60°
G 120° J 90°

11. Find the measure of ∠5.
A 30° C 60°
(B) 120° D 90°

12. Find the measure of ∠6.
F 30° (H) 60°
G 120° J 90°

13. Find the measure of ∠7.
A 45° C 60°
(B) 120° D 90°

LESSON 7-2
Reading Strategies
Understanding Symbols

The symbol ∥ stands for **parallel.** Parallel lines are the same distance apart and never meet.

The yard lines marked on a football field are similar to parallel line segments.

10 20 30 40 50 → A C E G I / B D F H J

Segment *AB* is parallel to segment *CD*. This can be written: $\overline{AB} \parallel \overline{CD}$.

1. Identify another pair of parallel line segments in the figure above.

Possible answer: $\overline{EF}$ is parallel to $\overline{GH}$.

2. Use the ∥ symbol to write how the line segments are related.

$\overline{EF} \parallel \overline{GH}$

The symbol ⊥ stands for **perpendicular.** Perpendicular lines meet to form a square corner.

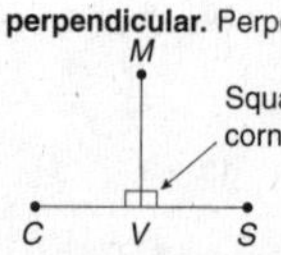

M, Square corner, C V S

MV ⊥ CS
Read: segment *MV* is perpendicular to segment *CS*.

A square in a corner is used as a symbol in a figure to show that lines are perpendicular.

Use this figure for Exercises 3–5.

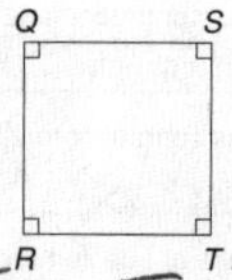

Q S / R T

3. Segment *QR* is perpendicular to segment *TR*. Write this with symbols.

$\overline{QR} \perp \overline{TR}$

4. Identify two other line segments that are perpendicular to each other.

Possible answer: $\overline{QS} \perp \overline{ST}$

5. What symbol in the figure tells you that line segments are perpendicular to each other?

the square in the corner

LESSON 7-2 **Puzzles, Twisters & Teasers**

Line Up!

Decide whether the lettered lines in each figure are parallel or perpendicular. Each answer has a corresponding letter. Circle the letter above your answer. Use the letters to solve the riddle.

1. (R) parallel — J perpendicular
2. U parallel — (E) perpendicular
3. (C) parallel — T perpendicular
4. P parallel — (C) perpendicular
5. (I) parallel — R perpendicular
6. O parallel — (N) perpendicular

What do you call it when you ride your bike around and around the block?

R E C Y C L I N G

1 2 3 4 5 6

LESSON 7-3 **Practice A**

Angles in Triangles

Identify each triangle by its angles and sides.

1. acute, isosceles
2. right, scalene
3. obtuse, isosceles

Find each angle measure.

4. Find $x°$ in the acute triangle. 65°
5. Find $y°$ in the right triangle. 40°
6. Find $r°$ in the obtuse triangle. 30°
7. Find $x°$ in the acute triangle. 30°
8. Find $y°$ in the right triangle. 42°
9. Find $m°$ in the obtuse triangle. 25°
10. Find $t°$ in the isosceles triangle. 70°
11. Find $x°$ in the scalene triangle. 40°
12. Find $n°$ in the isoceles triangle. 65°
13. The second angle in a triangle is three times as large as the first. The third angle is one third as large as the second angle. Find the angle measures. Draw a possible picture of the triangle.

36°, 108°, 36° — sample answer

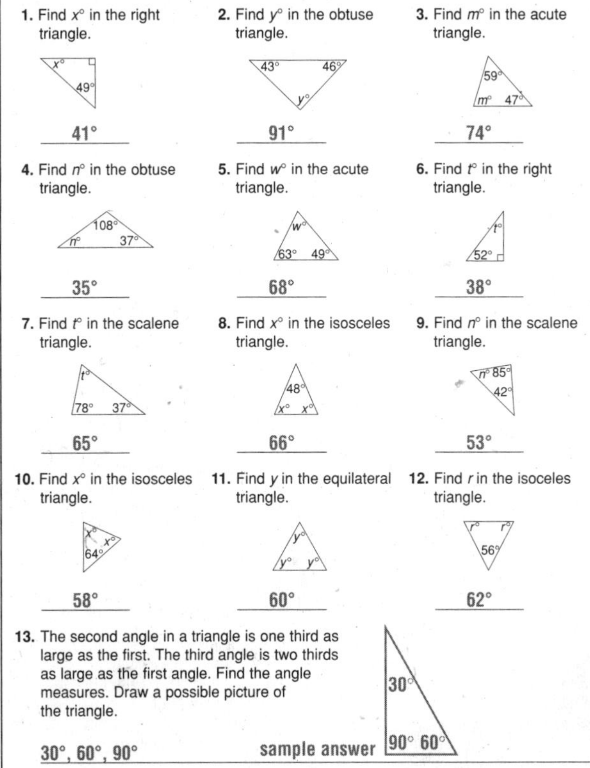

LESSON 7-3 **Practice B**

Angles in Triangles

1. Find $x°$ in the right triangle. 41°
2. Find $y°$ in the obtuse triangle. 91°
3. Find $m°$ in the acute triangle. 74°
4. Find $n°$ in the obtuse triangle. 35°
5. Find $w°$ in the acute triangle. 68°
6. Find $t°$ in the right triangle. 38°
7. Find $t°$ in the scalene triangle. 65°
8. Find $x°$ in the isosceles triangle. 66°
9. Find $n°$ in the scalene triangle. 53°
10. Find $x°$ in the isosceles triangle. 58°
11. Find y in the equilateral triangle. 60°
12. Find r in the isoceles triangle. 62°
13. The second angle in a triangle is one third as large as the first. The third angle is two thirds as large as the first angle. Find the angle measures. Draw a possible picture of the triangle.

30°, 60°, 90° — sample answer

LESSON 7-3 **Practice C**

Angles in Triangles

Find the value of each variable.

1. 59°
2. 46°
3. 60°
4. 31°
5. 57°
6. 30°
7. The measure of the second angle in a triangle is four more than the measure of the first angle and the measure of the third angle is eight more than twice the measure of the first angle. Find the measure of each angle.

first ∠ = 42°; second ∠ = 46°; third ∠ = 92°

Describe each statement as *always*, *sometimes*, or *never* true.

8. An obtuse triangle is a scalene triangle.

sometimes

9. A scalene triangle is an isosceles triangle.

never

10. A right triangle is an isosceles triangle.

sometimes

11. A triangle with all angles congruent is acute.

always

Sketch a triangle to fit each description. If no triangle can be drawn, write *not possible*. Possible answers:

12. obtuse, isosceles
13. right, scalene
14. acute, equilateral

LESSON 7-3

Reteach

Angles in Triangles

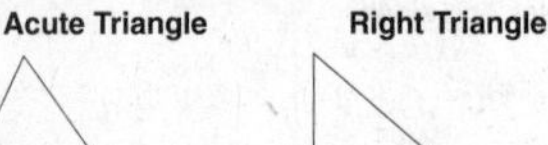

Acute Triangle | **Right Triangle** | **Obtuse Triangle**

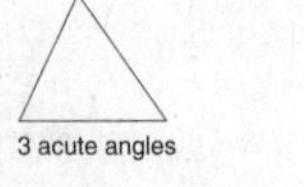

3 acute angles

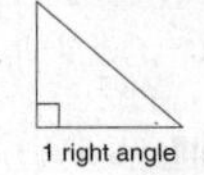

1 right angle

1 obtuse angle

Triangle Sum Theorem: The sum of the measures of the three interior angles of any triangle is 180°.

In the diagram: $m\angle A + m\angle B + m\angle C = 180°$

$a° + b° + c° = 180°$

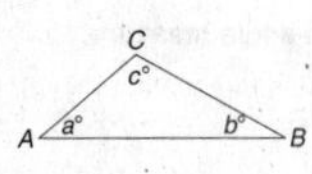

Complete to find the measure of the unknown angle.

1.

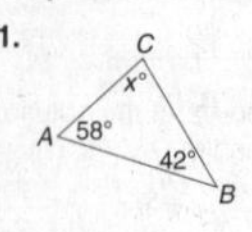

$m\angle A + m\angle B + m\angle C = 180°$

58° + 42° + $x°$ = 180°

100° + $x°$ = 180°

$x°$ = 80°

2.

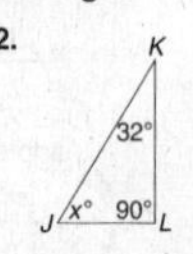

$m\angle J + m\angle K + m\angle L = 180°$

$x°$ + 32° + 90° = 180°

$x°$ + 122° = 180°

$x°$ = 58°

3.

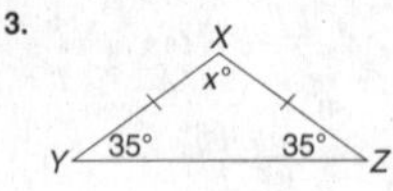

$m\angle X + m\angle Y + m\angle Z = 180°$

$x° + 35° + 35° = 180°$

$x° + 70° = 180°$

$x°$ = 110°

4.

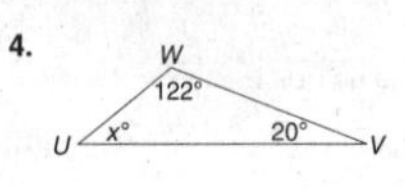

$m\angle U + m\angle V + m\angle W = 180°$

$x° + 20° + 122° = 180°$

$x° + 142° = 180°$

$x°$ = 38°

LESSON 7-3

Challenge

Change a This into a That

A **geometric dissection** involves cutting a figure into pieces that can then be rearranged to form another figure.

Trace each figure. Cut up the figure you have traced and rearrange the numbered pieces to form the indicated figure. Sketch your solution.

1. Rearrange the pieces of the equilateral triangle to form a square.

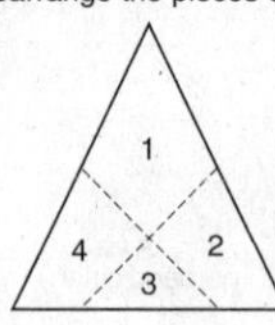

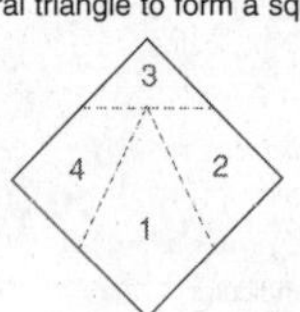

2. Rearrange the pieces of the star to form an equilateral triangle.

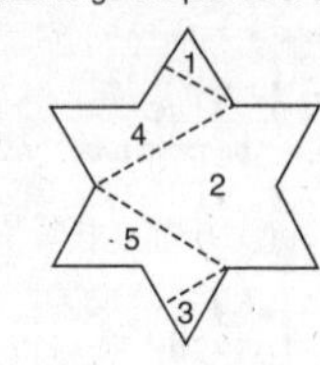

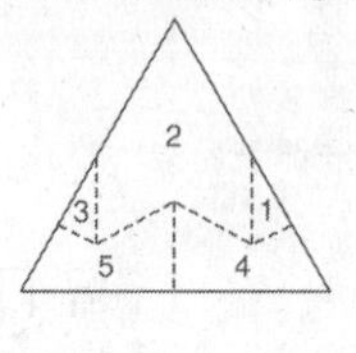

3. Rearrange the pieces of the cross to form an equilateral triangle.

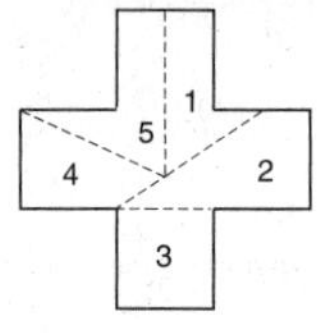

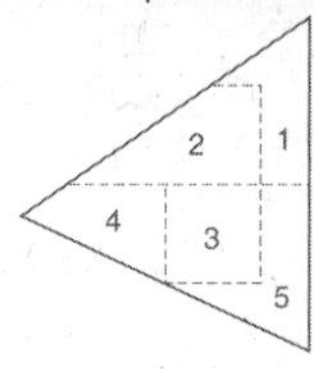

LESSON 7-3

Problem Solving

Angles in Triangles

The American flag must be folded according to certain rules that result in the flag being folded into the shape of a triangle. The figure shows a frame designed to hold an American flag.

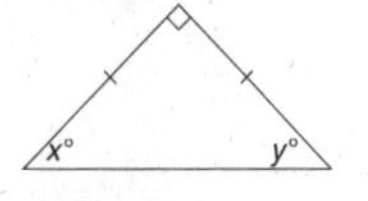

1. Is the triangle acute, right, or obtuse?

right

2. Is the triangle equilateral, isosceles, or scalene?

isosceles

3. Find $x°$.

$x = 45°$

4. Find $y°$.

$y = 45°$

The figure shows a map of three streets. Choose the letter for the best answer.

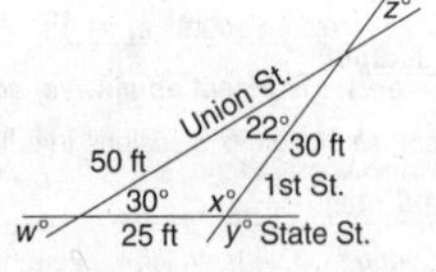

5. Find $x°$.
   - A 22°
   - (B) 128°
   - C 30°
   - D 68°

6. Find $w°$.
   - F 22°
   - G 128°
   - (H) 30°
   - J 52°

7. Find $y°$.
   - A 22°
   - B 30°
   - (C) 128°
   - D 143°

8. Find $z°$.
   - (F) 22°
   - G 30°
   - H 128°
   - J 143°

9. Which word best describes the triangle formed by the streets?
   - A acute
   - B right
   - (C) obtuse
   - D equilateral

10. Which word best describes the triangle formed by the streets?
    - F equilateral
    - G isosceles
    - (H) scalene
    - J acute

LESSON 7-3

Reading Strategies

Graphic Organizer

A triangle can be classified by the measurement of one or more of its angles.

Right angle = 90° Acute angle < 90° Obtuse angle > 90°

A triangle can also be classified by the length of its sides. This chart will help you compare triangles by angles and sides.

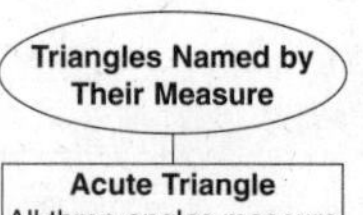

Triangles Named by Their Measure	Triangles Named by Length of Sides
Acute Triangle All three angles measure less than 90°.	**Equilateral Triangle** All three sides are equal.
Right Triangle One angle measures 90°.	**Isosceles Triangle** Two sides are equal.
Obtuse Triangle One angle measures more than 90°.	**Scalene Triangle** None of the sides are equal.

Write two of the following terms to describe each triangle: acute, right, obtuse, equilateral, isosceles, scalene.

1.

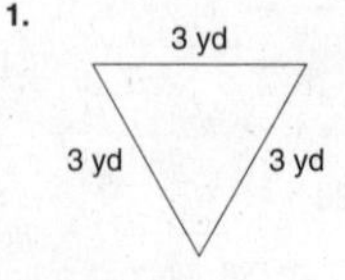

equilateral, acute

2.

6.5 in. 4 in. 4 in.

right, isosceles

LESSON 7-3

Puzzles, Twisters & Teasers

Try Some Triangles

Across

1. A ______ triangle has one 90° angle.
5. An ______ triangle has 3 acute angles.
6. The Triangle Sum ______ says that the 3 angles in a triangle always equal 180°.
7. A ______ triangle has no congruent sides and no congruent angles.

Down

2. An ______ triangle has at least 2 congruent sides and 2 congruent angles.
3. An ______ triangle has 3 congruent sides and 3 congruent angles.
4. An ______ triangle has 1 obtuse angle.

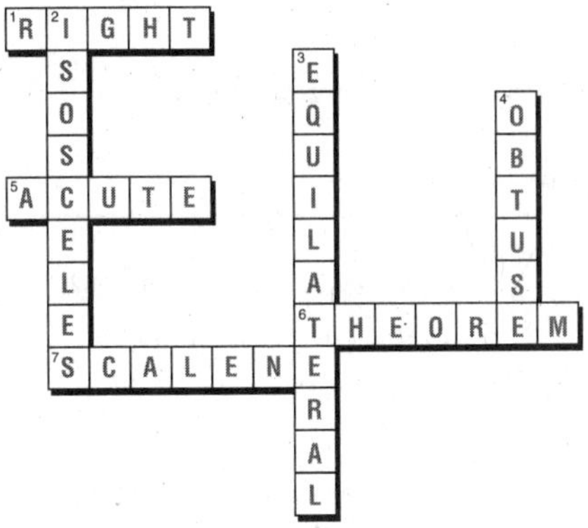

Answers: 1. RIGHT; 2. ISOSCELES; 3. EQUILATERAL; 4. OBTUSE; 5. ACUTE; 6. THEOREM; 7. SCALENE

LESSON 7-4

Practice A

Classifying Polygons

Name each polygon.

1.  pentagon
2.  octagon
3.  hexagon

Find the sum of the angle measures in each figure.

4. 360°
5.  180°
6.  540°
7.  360°
8.  720°
9. 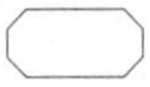 1080°

Find the angle measures in each regular polygon.

10. 60°
11.  144°
12. 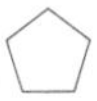 108°
13. 135°
14. 90°
15.  120°

Write all the names that apply to each figure.

16.  quadrilateral, parallelogram, rectangle
17. 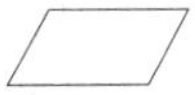 quadrilateral, parallelogram
18. quadrilateral, parallelogram, rectangle, rhombus, square

LESSON 7-4

Practice B

Classifying Polygons

Find the sum of the angle measures in each figure.

1.  360°
2.  720°
3.  900°
4. 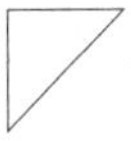 180°
5. 1080°
6.  540°

Find the angle measures in each regular polygon.

7. 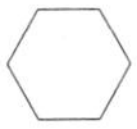 120°
8. 90°
9. 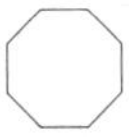 135°
10. 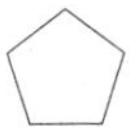 108°
11.  60°
12. 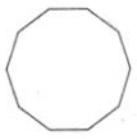 144°

Give all the names that apply to each figure.

13.  quadrilateral, parallelogram, rhombus
14. 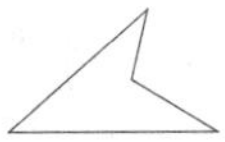 quadrilateral
15. 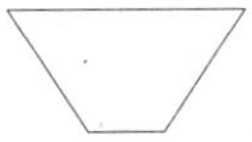 quadrilateral, trapezoid

LESSON 7-4

Practice C

Classifying Polygons

Find the sum of the angle measures in each regular polygon. Then, find the measure of each angle.

1. 24-gon — 3960°; 165°
2. 16-gon — 2520°; 157.5°
3. 36-gon — 6120°; 170°

Find the value of each variable.

4. 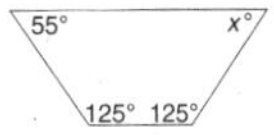 55° (x°; 125° 125°) — 55°
5. 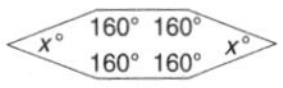 (x°, 160° 160°, 160° 160°, x°) — 40°
6. 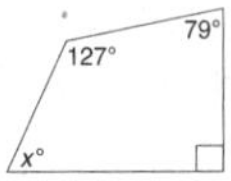 (127°, 79°, x°) — 64°

The sum of the angle measures of a polygon is given. Name the polygon.

7. 1080° — octagon
8. 1260° — nonagon
9. 900° — heptagon
10. 1440° — decagon

Graph the given vertices on a coordinate plane. Connect the points to draw a polygon and classify it by the number of its sides.

11. (1, 5), (4, 2), (4, −2), (1, −5), (−3, −5), (−5, −2), (−5, 2), (−3, 5) — octagon-eight sides
12. (0, −1), (−1, 3), (2, 5), (5, 3), (4, −1) — pentagon-five sides

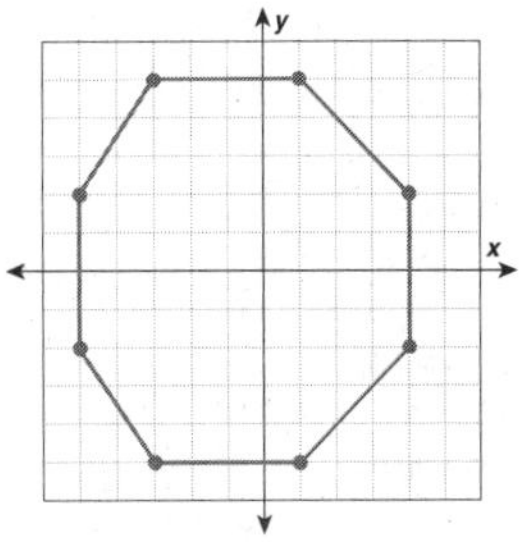

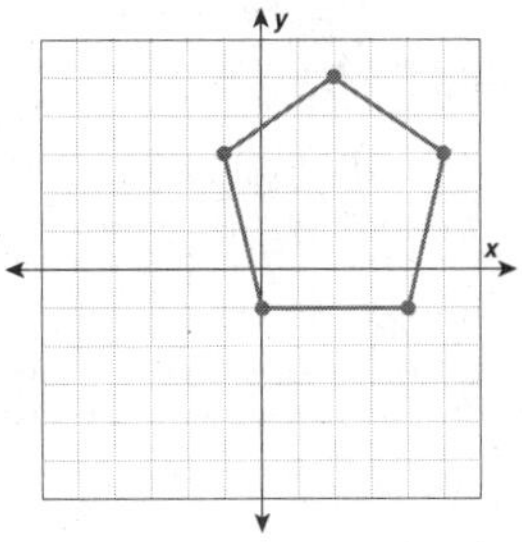

LESSON 7-4 **Reteach**

Classifying Polygons

A polygon of n sides (an **n-gon**) can be divided into $(n - 2)$ triangles
The sum of the angle measures of an n-gon $= (n - 2)180°$.

A polygon of 5 sides (pentagon) can be divided into 3 triangles.

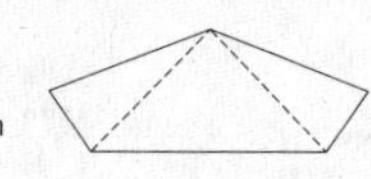

Sum of angle measures of pentagon
$= (n - 2)\ 180°$
$= (5 - 2)\ 180° = (3)180° = 540°$

Find the sum of the measures of the angles.

	1. quadrilateral	2. hexagon
How many sides in the polygon?	4	6
How many triangles can be formed?	4 − 2 = 2	6 − 2 = 4
Multiply the number of triangles by 180°.	180° × 2	180° × 4
sum of the measures of the angles	360°	720°

In a **regular polygon,** all sides and all angles are congruent.
The measure of each angle of a regular polygon $= \frac{\text{sum of the angles}}{\text{number of sides}}$

The measure of each angle of a regular pentagon $= \frac{(5-2)180°}{5} = \frac{(3)180°}{5} = \frac{540°}{5} = 108°$

Find the measure of each angle.

	3. regular octagon	4. regular decagon
How many sides (angles) in the polygon?	8	10
How many triangles can be formed?	8 − 2 = 6	10 − 2 = 8
Multiply the number of triangles by 180°.	180° × 6	180° × 8
Sum of the measures of the angles	1080°	1440°
Divide the sum by the number of angles.	$\frac{1080°}{8}$	$\frac{1440°}{10}$
Measure of each angle of the polygon	135°	144°

LESSON 7-4 **Reteach**

Classifying Polygons (continued)

Quadrilaterals 4 sides, 4 angles
- **Parallelograms** 2 pairs of parallel sides
  - **Rectangles** 4 right angles
  - **Rhombuses** 4 congruent sides
    - **Squares** 4 right angles, 4 congruent sides
- **Trapezoids** exactly 1 pair of parallel sides

A figure on a lower branch of the tree has the properties of the figures above it.

All rectangles are parallelograms.
But not all parallelograms are rectangles.

Write all the names that apply to each figure.

5. $\overline{AB} \parallel \overline{DC}$
   - four-sided polygon — quadrilateral
   - 1 pair of parallel sides — trapezoid

6. $\overline{WX} \parallel \overline{ZY}, \overline{WZ} \parallel \overline{XY}$; $\overline{WX} \cong \overline{XY} \cong \overline{ZY} \cong \overline{WZ}$
   - four-sided polygon — quadrilateral
   - 2 pairs of parallel sides — parallelogram
   - 4 congruent sides — rhombus

7. $\overline{RS} \parallel \overline{UT}, \overline{RU} \parallel \overline{ST}$
   - four-sided polygon — quadrilateral
   - 2 pairs of parallel sides — parallelogram
   - 4 right angles — rectangle

LESSON 7-4 **Challenge**

Slanted View

1. Refer to parallelogram $ABCD$. Use a ruler.
   a. Is diagonal $\overline{AC} \cong$ diagonal $\overline{BD}$? no
   b. Is $\overline{AM} \cong \overline{MC}$? Is $\overline{DM} \cong \overline{MB}$? yes, yes

   Make a statement about how the diagonals of a parallelogram relate to each other.

   The diagonals of a parallelogram are not congruent but they do bisect each other.

2. Refer to rectangle $ABCD$ and your observations in Question 1.
   a. Since a rectangle is a parallelogram, what property should the diagonals have? Use a ruler to verify your conjecture.

   Diagonals bisect each other.

   b. What additional property do the diagonals of a rectangle have?

   Diagonals are congruent.

3. Refer to rhombus $ABCD$.
   a. Since a rhombus is a parallelogram, what property should the diagonals have? Use a ruler to verify your conjecture.

   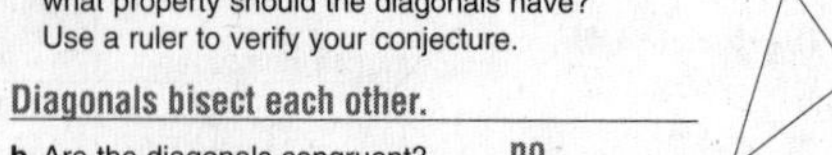

   Diagonals bisect each other.

   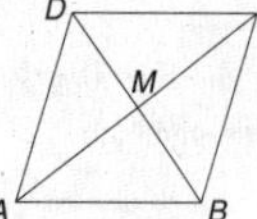

   b. Are the diagonals congruent? no
   c. Measure the angles with vertex M. What additional property do the diagonals of a rhombus have?

   Diagonals are perpendicular to each other.

   d. Measure the angles at each vertex of the rhombus. What additional property do the diagonals of a rhombus have?

   Diagonals bisect the opposite angles.

4. Make a conjecture about the properties of the diagonals of a square. Draw a square and verify your conjectures with a ruler and protractor.

   congruent; perpendicular bisectors of each other; bisect opposite angles

LESSON 7-4 **Problem Solving**

Classifying Polygons

The figure shows how the glass for a window will be cut from a square piece. Cuts will be made along $\overline{CE}$, $\overline{FH}$, $\overline{IK}$, and $\overline{LB}$.

1. What shape is the window?

   octagon

2. What is the sum of the angle measures of the window?

   1080°

3. What is the measure of each angle of the window?

   135°

4. Based on the angles, what kind of triangle is $\triangle CDE$?

   right triangle

5. Based on the sides, what kind of triangle is $\triangle CDE$?

   isosceles triangle

The figure shows how parallel cuts will be made along $\overline{AD}$ and $\overline{BC}$. $\overline{AB}$ and $\overline{CD}$ are parallel. Choose the letter for the best answer.

6. Which word correctly describes figure $ABCD$ after the cuts are made?
   A triangle
   (B) quadrilateral
   C pentagon
   D hexagon

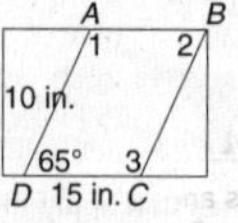

7. Which word correctly describes figure $ABCD$ after the cuts are made?
   (F) parallelogram
   G trapezoid
   H rectangle
   J rhombus

8. Find the measure of $\angle 1$.
   A 45°
   B 65°
   C 90°
   (D) 115°

9. Find the measure of $\angle 2$.
   F 45°
   G 90°
   (H) 65°
   J 115°

10. Find the measure of $\angle 3$.
    A 45°
    B 90°
    C 65°
    (D) 115°

LESSON 7-4 Reading Strategies
Reading a Chart

The prefix *poly-* means "many." The word *polygon* means "many-sided figure." A **polygon** is a closed, plane figure named by the number of its sides.

Polygons **Not Polygons**

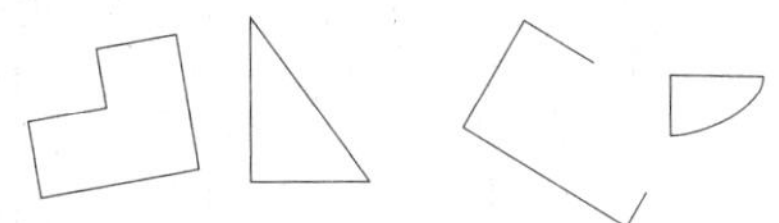

The prefix of a polygon's name tells you the number of its sides and angles.

- *Tri-* means "three" → A **tri**cycle has 3 wheels.
  → A **tri**angle has 3 sides and 3 angles.
- *Quad-* means "four" → A **quad**ruped is a 4-legged animal.
  → A **quad**rilateral has 4 sides and 4 angles.

This chart helps you organize polygons by their sides and angles.

Polygon	Number of Sides	Number of Angles
Triangle	3	3
Quadrilateral	4	4
Pentagon	5	5
Hexagon	6	6
Octagon	8	8

Write the name of each polygon. Use the chart to help you.

1. 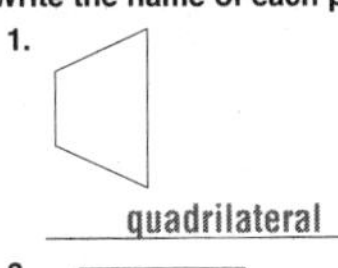
quadrilateral

2. 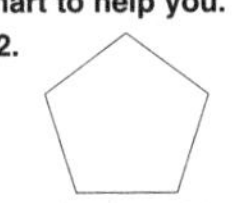
pentagon

3. hexagon

4. 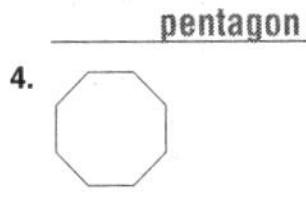
octagon

LESSON 7-4 Puzzles, Twisters & Teasers
What Side Are You On?!

Name each figure, one letter on each space. Each answer has one or two boxed letters. Unscramble the boxed letters to solve the riddle.

1. 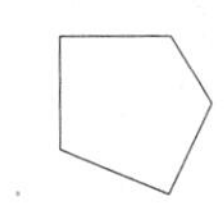
P E N T [A] G O N

2. O C T [A] G O N

3. 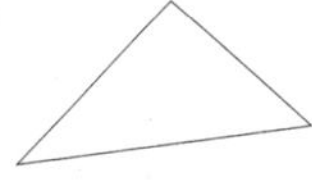
T R [I] A N G L E

4. 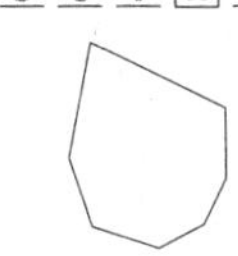
[H] E P T A G O N

5. 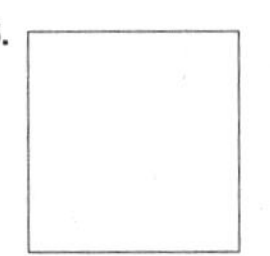
Q U A [D] R I L A T E [R] A L

6. 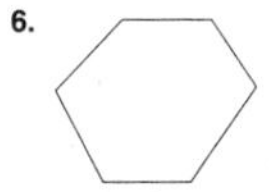
H [E] X A G O N

Why was the computer tired after its long trip?

It was A H A R D D R I V E.

LESSON 7-5 Practice A
Coordinate Geometry

Fill in each blank with the correct word from the box at the right.

downward horizontal upward vertical

1. If a line has a positive slope, it slants upward to the right.
2. If a line has a negative slope, it slants downward to the right.
3. The slope of a vertical line is undefined.
4. A horizontal line has a slope of 0.

Determine if the slope of each line is positive, negative, 0, or undefined. Then find the slope of each line.

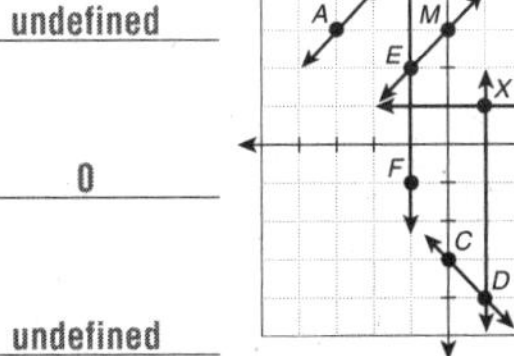

5. $\overleftrightarrow{AB}$ positive; 1
6. $\overleftrightarrow{EF}$ undefined
7. $\overleftrightarrow{CD}$ negative; −1
8. $\overleftrightarrow{XY}$ 0
9. $\overleftrightarrow{EM}$ positive; 1
10. $\overleftrightarrow{DX}$ undefined

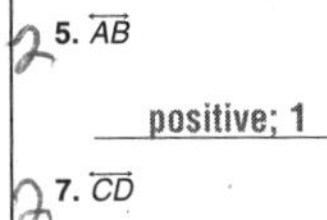

11. Which lines are parallel?
$\overleftrightarrow{AB} \parallel \overleftrightarrow{EM}$; $\overleftrightarrow{EF} \parallel \overleftrightarrow{DX}$
12. Which lines are perpendicular?
$\overleftrightarrow{AB} \perp \overleftrightarrow{CD}$; $\overleftrightarrow{EM} \perp \overleftrightarrow{CD}$; $\overleftrightarrow{EF} \perp \overleftrightarrow{XY}$; $\overleftrightarrow{DX} \perp \overleftrightarrow{XY}$

Graph the quadrilateral with the given vertices. Write all the names that apply to the quadrilateral.

13. (−2, 2), (3, 2), (1, −2), (−4, −2)
quadrilateral
parallelogram

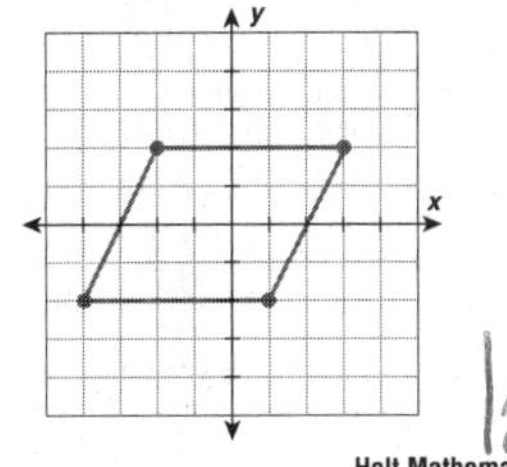

LESSON 7-5 Practice B
Coordinate Geometry

Determine if the slope of each line is positive, negative, 0, or undefined. Then find the slope of each line.

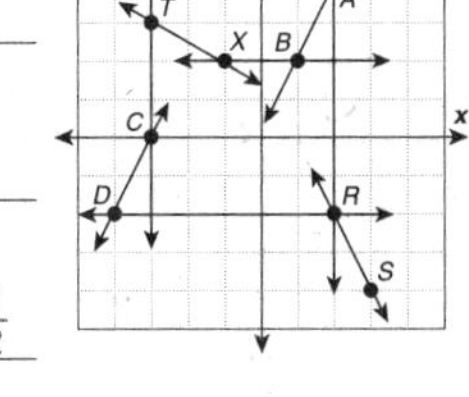

1. $\overleftrightarrow{AB}$ positive; 2
2. $\overleftrightarrow{CD}$ positive; 2
3. $\overleftrightarrow{RS}$ negative; −2
4. $\overleftrightarrow{TC}$ undefined
5. $\overleftrightarrow{DR}$ 0
6. $\overleftrightarrow{TX}$ negative; $-\frac{1}{2}$
7. Which lines are parallel?
$\overleftrightarrow{AB} \parallel \overleftrightarrow{CD}$; $\overleftrightarrow{TC} \parallel \overleftrightarrow{AR}$; $\overleftrightarrow{XB} \parallel \overleftrightarrow{DR}$
8. Which lines are perpendicular?
$\overleftrightarrow{AB} \perp \overleftrightarrow{TX}$; $\overleftrightarrow{CD} \perp \overleftrightarrow{TX}$; $\overleftrightarrow{DR} \perp \overleftrightarrow{TC}$; $\overleftrightarrow{DR} \perp \overleftrightarrow{AR}$; $\overleftrightarrow{TC} \perp \overleftrightarrow{XB}$; $\overleftrightarrow{AR} \perp \overleftrightarrow{XB}$

Graph the quadrilateral with the given vertices. Write all the names that apply to the quadrilateral.

9. (−1, 1), (4, 1), (1, −3), (−4, −3)

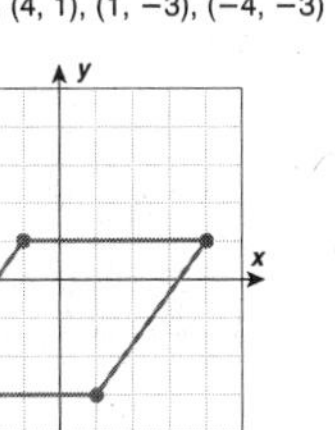

quadrilateral, parallelogram, rhombus

Find the coordinates of the missing vertex.

10. rhombus *ABCD* with *A*(0, 4), *B*(4, 1), and *C*(0, −2)

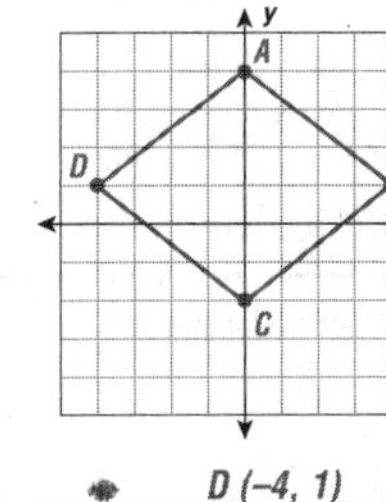

D (−4, 1)

LESSON 7-5

Practice C

Coordinate Geometry

Draw the line through the given points and find its slope.

1. $A(3, 2)$, $B(4, 4)$ 2
2. $C(-2, 1)$, $D(-2, 3)$ undefined
3. $R(-1, -4)$, $S(-3, 4)$ -4
4. $X(2, -3)$, $Y(4, -4)$ $-\frac{1}{2}$
5. $M(-3, -3)$, $N(0, -2)$ $\frac{1}{3}$
6. $E(2, 3)$, $F(0, 1)$ 1

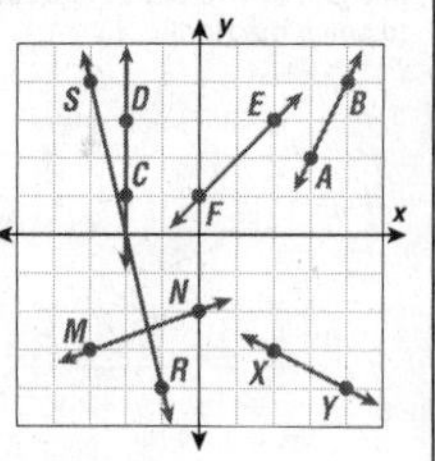

Find the coordinates of the missing vertex.

7. parallelogram $RSTU$ with $R(-4, 4)$, $S(2, 4)$, $T(4, 0)$

$U(-2, 0)$

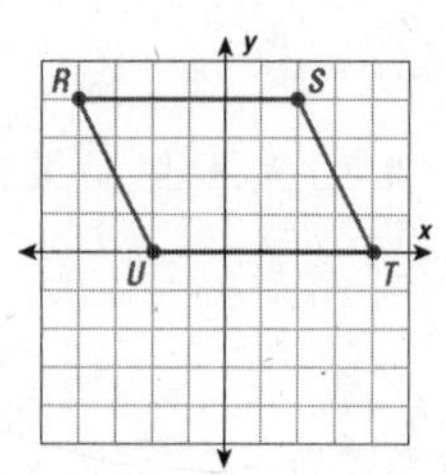

8. On a coordinate grid draw a line r with a slope 0 and a line s with slope 1. Then draw a line through the intersection of lines r and s that has a slope between 0 and 1. Name the line and state its slope.

sample answer given:
line a with slope $= \frac{1}{4}$;

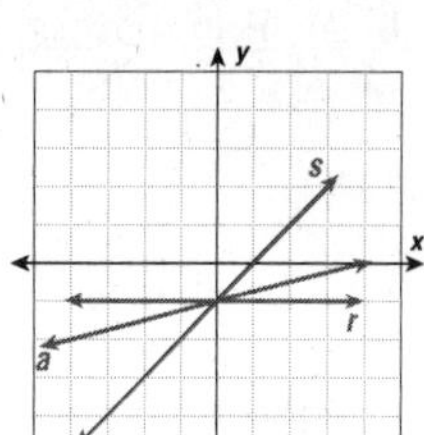

LESSON 7-5

Reteach

Coordinate Geometry

Possible Values for Slope

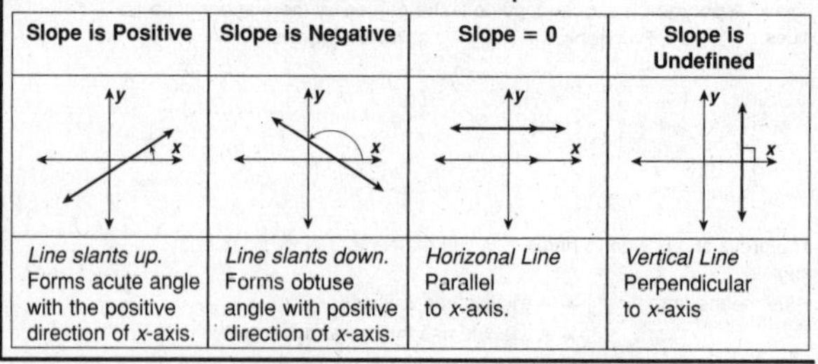

Slope is Positive	Slope is Negative	Slope = 0	Slope is Undefined
Line slants up. Forms acute angle with the positive direction of x-axis.	*Line slants down.* Forms obtuse angle with positive direction of x-axis.	*Horizonal Line* Parallel to x-axis.	*Vertical Line* Perpendicular to x-axis

Plot the given points. Describe the slope of the line that joins them.

1. $(-2, 2)$ and $(2, 5)$

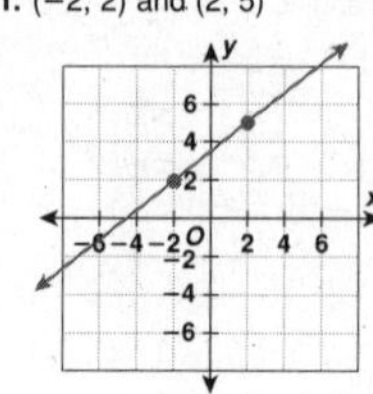

slope is: positive

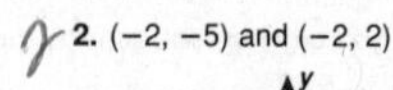

2. $(-2, -5)$ and $(-2, 2)$

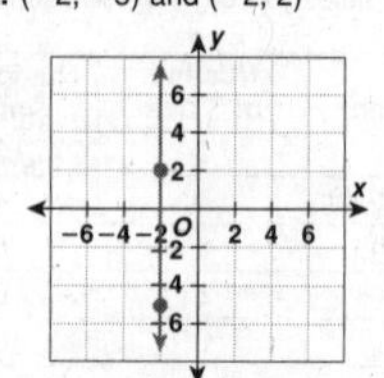

slope is: undefined

3. $(1, 2)$ and $(5, -2)$

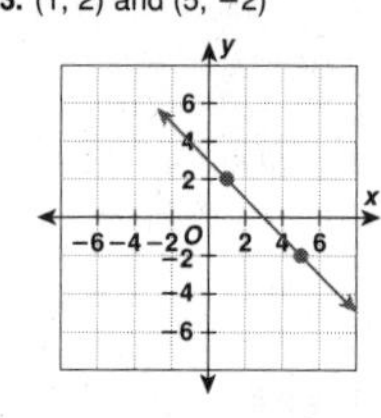

slope is: negative

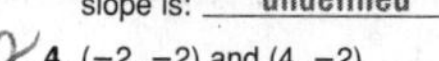

4. $(-2, -2)$ and $(4, -2)$

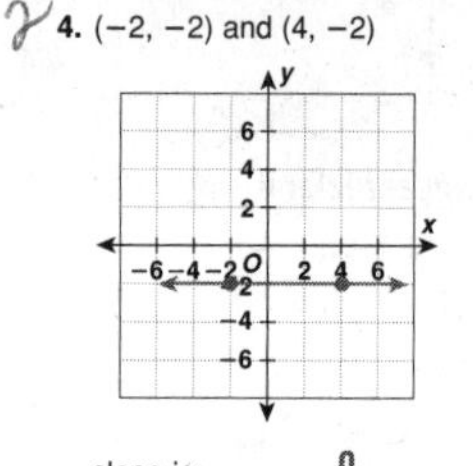

slope is: 0

LESSON 7-5

Reteach

Coordinate Geometry (continued)

To find the slope of a line, use a *direction ratio* such as $\frac{\text{up}}{\text{right}}$.

direction ratio from A to $B = \frac{\text{up } 5}{\text{right } 3}$

slope of $\overline{AB} = \frac{5}{3}$

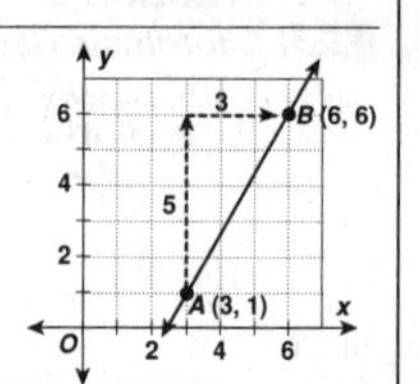

Complete to find the slope of each line.

5.

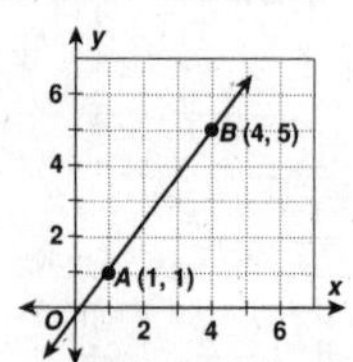

From A to B, do you go up or down? How many units? up 4

Do you go right or left? How many units? right 3

slope of $\overline{AB} = \frac{4}{3}$

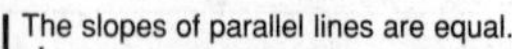

The slopes of parallel lines are equal.

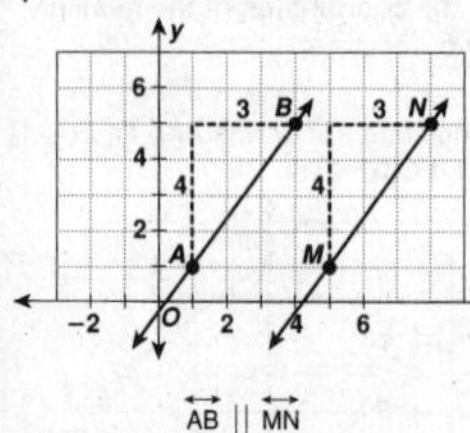

$\overleftrightarrow{AB} \parallel \overleftrightarrow{MN}$

The product of the slopes of perpendicular lines is -1.

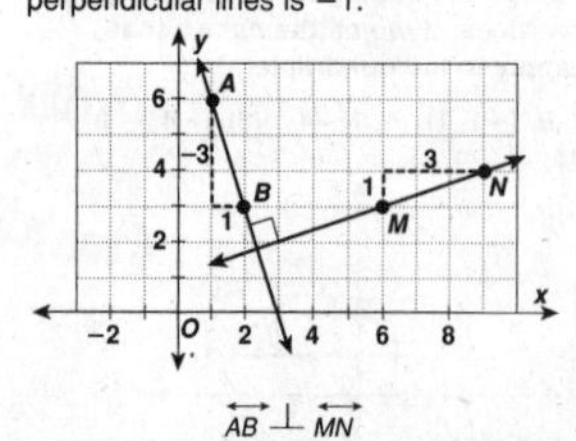

$\overleftrightarrow{AB} \perp \overleftrightarrow{MN}$

Complete each statement. If the slope of $\overline{CD} = -\frac{2}{3}$

6. and $\overline{CD}$ is parallel to $\overline{XY}$, then the slope of $\overline{XY}$ is: $-\frac{2}{3}$
7. and $\overline{CD}$ is perpendicular to $\overline{PQ}$, then the slope of $\overline{PQ}$ is: $\frac{3}{2}$

LESSON 7-5

Challenge

Are They Lined Up?

You can find the slope of a line by using the coordinates of two points on the line.

$$\text{slope} = \frac{\text{difference of } y\text{-values}}{\text{difference of } x\text{-values}}$$

Be sure to take the differences in the same order.

To find the slope of $\overline{AB}$ with $A(-2, 5)$ and $B(6, 7)$:

slope of $\overline{AB} = \frac{7 - 5}{6 - (-2)} = \frac{2}{8} = \frac{1}{4}$, or

slope of $\overline{AB} = \frac{5 - 7}{-2 - 6} = \frac{-2}{-8} = \frac{1}{4}$

Find the slope of the line joining each pair of points. Verify your result on a graph.

1. $(5, 5)$ and $(2, 1)$ $\frac{4}{3}$

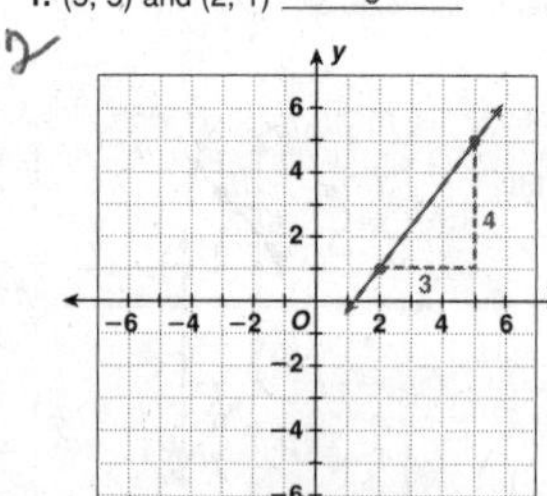

2. $(4, 1)$ and $(6, -2)$ $-\frac{3}{2}$

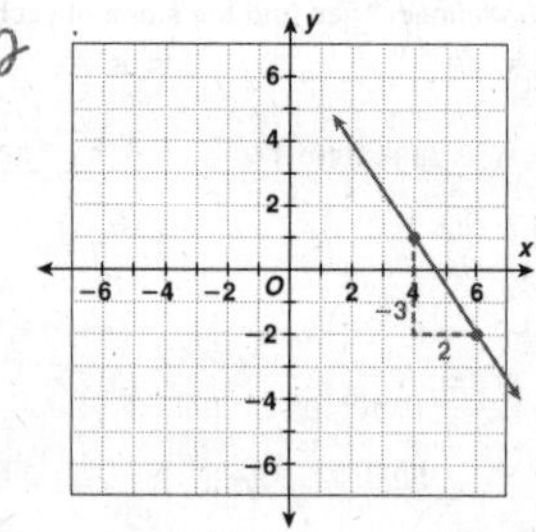

3. Find the value of k so that the slope of the line joining the points $(k, -3)$ and $(4, 2)$ is $\frac{1}{2}$. $k =$ -6

The slope between any two points on a line is the same everywhere on that line; that is, the slope of a given line is *constant.*

Without a graph, determine if each set of points is collinear (lie on the same line). Explain your method.

4. $A(0, -4)$, $B(1, -2)$, and $C(3, 2)$

slope $\overline{AB} = 2$; slope $\overline{BC} = 2$;
yes

5. $P(-7, -1)$, $Q(1, 7)$, and $R(7, 1)$

slope $\overline{PQ} = 1$; slope $\overline{QR} = -1$;
no

6. Find the value of k so that the points $L(-1, 5)$, $M(0, k)$ and $N(1, -1)$ are collinear. $k =$ 2

LESSON **7-5**

Problem Solving

Coordinate Geometry

The Uniform Federal Accessibility Standards describes the standards for making buildings accessible for the handicapped. The standards say that the least possible slope should be used for a ramp and that the maximum slope of a ramp should be $\frac{1}{12}$.

1. What is the slope of the pictured ramp? Does the ramp meet the standard?

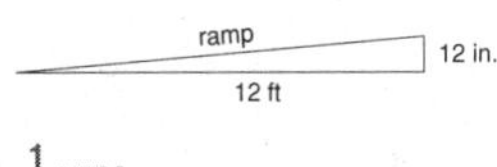

$\frac{1}{12}$; yes

2. What is the slope of the pictured ramp? Does the ramp meet the standard?

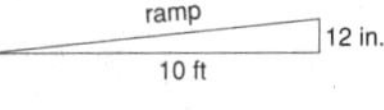

$\frac{1}{10}$; no

Write the correct answer.

3. Find the slope of the roof.

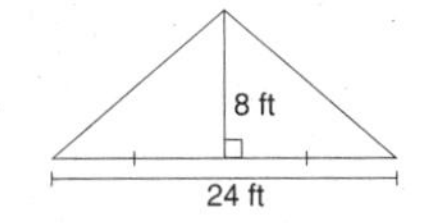

Slope $= \frac{2}{3}$ or $-\frac{2}{3}$

Choose the letter that represents the slope.

4. Many building codes require that a staircase be built with a maximum rise of 8.25 inches for a minimum tread width (run) of 9 inches.

A $\frac{8}{9}$ C $\frac{9}{8.25}$

(B) $\frac{11}{12}$ D $\frac{12}{11}$

5. Hills that have a rise of about 10 feet for every 17 feet horizontally are too steep for most cars.

(F) $\frac{10}{17}$ H $\frac{17}{10}$

G $\frac{2}{5}$ J $\frac{3}{5}$

6. At its steepest part, an intermediate ski run has a rise of about 4 feet for 10 feet horizontally.

(A) $\frac{2}{5}$ C $\frac{5}{2}$

B $\frac{4}{5}$ D $\frac{5}{4}$

7. Black diamond, or expert, ski slopes often have a rise of 10 feet for every 14 feet horizontally.

F $\frac{7}{5}$ (H) $\frac{5}{7}$

G $\frac{2}{7}$ J $\frac{7}{2}$

LESSON **7-5**

Reading Strategies

Using Graphic Aids

When a horizontal number line crosses a vertical number line, a **coordinate plane,** or grid, is formed. The point where the two lines meet is called the **origin,** or (0, 0).

Any location on the coordinate plane can be shown by a point. The pair of numbers that name a point on the coordinate plane is called an **ordered pair.**

Two numbers are needed to identify the location of a point in a coordinate plane.

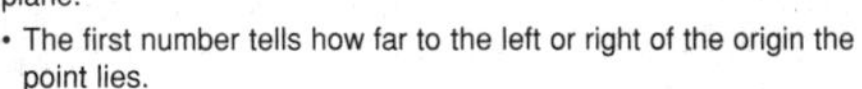

- The first number tells how far to the left or right of the origin the point lies.
- The second number tells how far up or down from the origin the point lies.

The location of point *M*, starting at (0, 0), is (–3, 1)→left 3, up 1.

1. Name the ordered pair for point *N*.

(–2, 4)

2. What ordered pair names point *D?*

(6, –3)

The **slope** is the slant of a line or line segment. Some slopes are steeper than others. Segment *TD* has no slope.

3. Which line segment on the coordinate plane looks like it has the steepest slope?

segment *MN*

4. Of line segments *MN*, *AB*, and *RQ*, which has the least amount of slope?

segment *RQ*

LESSON **7-5**

Puzzles, Twisters & Teasers

How Coordinated Are You?

Determine the slope of each line. In the letter box find the letter that matches each slope. Use the letter to fill in the blanks and solve the riddle.

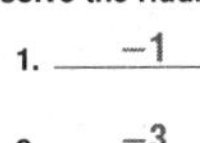

1. –1
2. –3
3. 2
4. 0
5. $\frac{1}{2}$

Letter Box

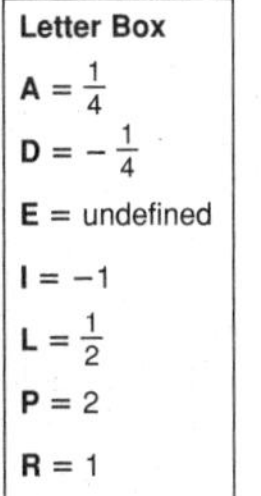

A = $\frac{1}{4}$
D = $-\frac{1}{4}$
E = undefined
I = –1
L = $\frac{1}{2}$
P = 2
R = 1
S = 0
T = $\frac{1}{3}$
U = –3
W = 4

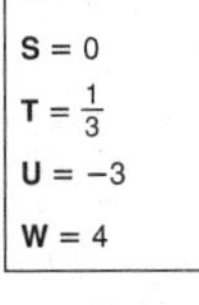

Why did the teacher go to the eye doctor?

She couldn't control her

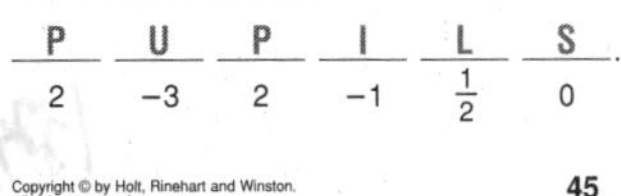

P	U	P	I	L	S
2	–3	2	–1	$\frac{1}{2}$	0

LESSON **7-6**

Practice A

Congruence

Match each polygon in column A with a congruent polygon in column B.

Column A / **Column B**

1. 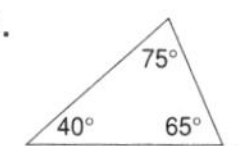

B

2. 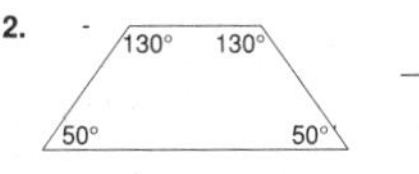

D

3. 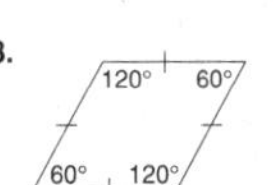

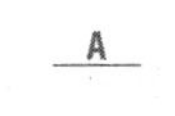
A

4. 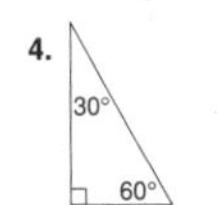

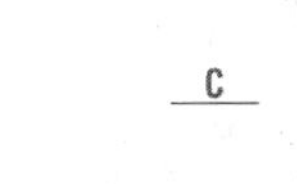
C

A. 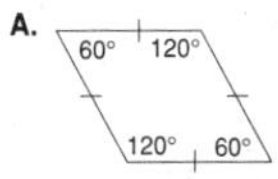

B. 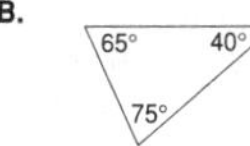

C. 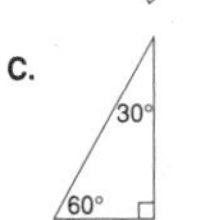

D. 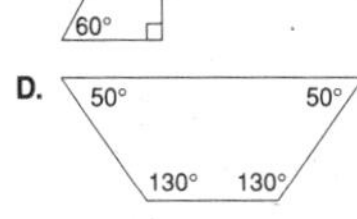

Write a congruence statement.

5. 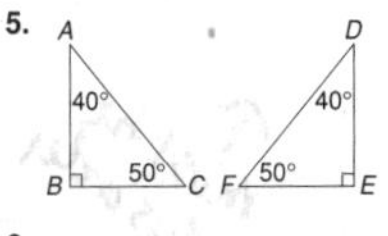

triangle *ABC* ≅ triangle *DEF*

6. 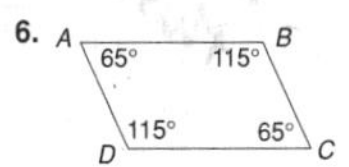

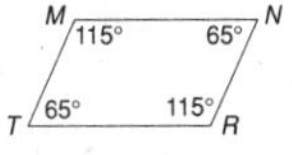

parallelogram *ABCD* ≅ parallelogram *TRNM*

Triangle *ABC* is congruent to triangle *WXY*.

7. Find *d*. 10   **8.** Find *t*. 55°

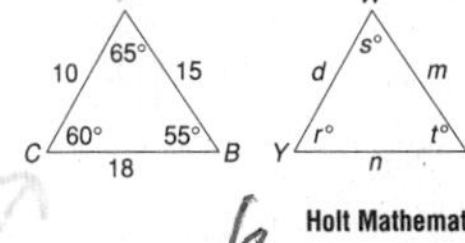

LESSON 7-6
Practice B
Congruence

Write a congruence statement for each pair of polygons.

1. 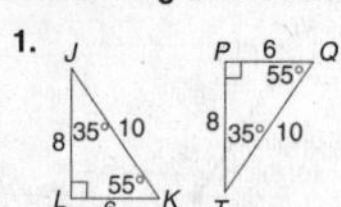

triangle $JKL \cong$ triangle TQP

2. 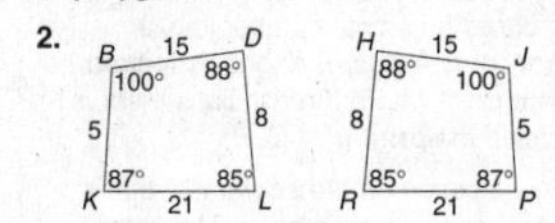

quadrilateral $BDLK \cong$ quadrilateral $JHRP$

3. 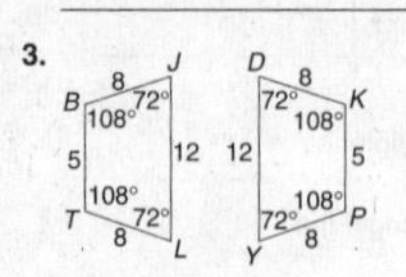

trapezoid $BJLT \cong$ trapezoid $KDYP$

4. 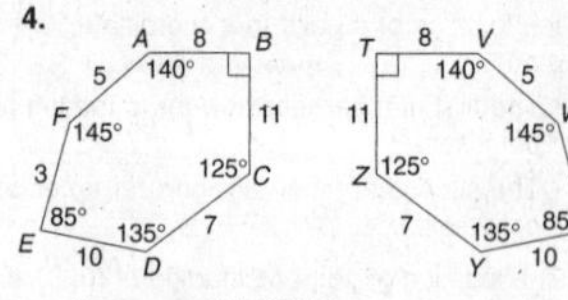

hexagon $ABCDEF \cong$ hexagon $VTZYXW$

In the figure, triangle $PRT \cong$ triangle FJH.

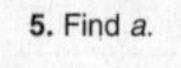
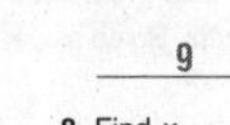
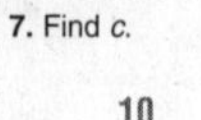
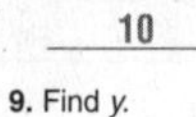

5. Find a. 4
6. Find b. 9
7. Find c. 10
8. Find x. 10°
9. Find y. 105°
10. Find z. 15°

LESSON 7-6
Practice C
Congruence

In the figure, quadrilateral $ABCD \cong$ quadrilateral $YZWX$.

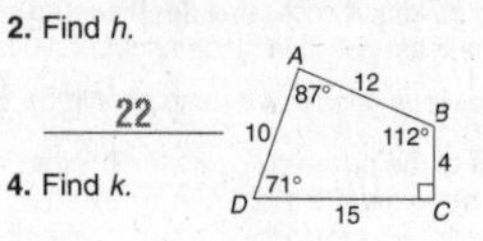

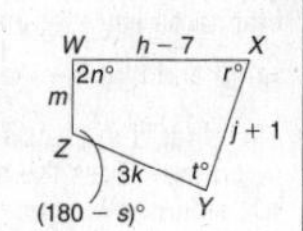

1. Find m. 4
2. Find h. 22
3. Find j. 9
4. Find k. 4
5. Find n. 45°
6. Find s. 68°
7. Find t. 87°
8. Find r. 71°

Find the value of the variables if octagon $ABCDEFGH$ is congruent to octagon $VWXYZSTU$.

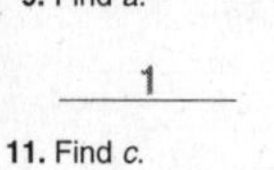
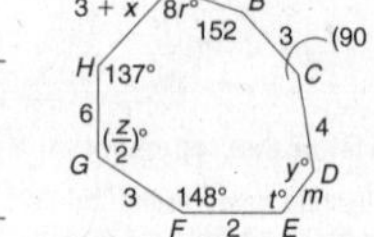

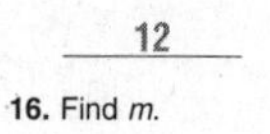
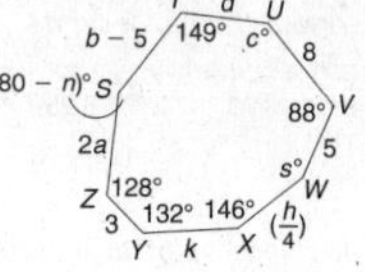

9. Find a. 1
10. Find b. 8
11. Find c. 137°
12. Find d. 6
13. Find g. 56°
14. Find h. 12
15. Find k. 4
16. Find m. 3
17. Find n. 32°
18. Find r. 11°
19. Find s. 152°
20. Find t. 128°
21. Find w. 4
22. Find x. 5
23. Find y. 132°
24. Find z. 298°

LESSON 7-6
Reteach
Congruence

Congruent polygons have the same size and shape.

Corresponding angles are congruent.

$\angle J \cong \angle J'$ $\angle K \cong \angle K'$ $\angle L \cong \angle L'$

(Read J' as J *prime*.)

Corresponding sides are congruent.

$\overline{JK} \cong \overline{J'K'}$ $\overline{KL} \cong \overline{K'L'}$ $\overline{LJ} \cong \overline{L'J'}$

In a congruence statement, the vertices of the second polygon are written in order of correspondence with the first polygon.

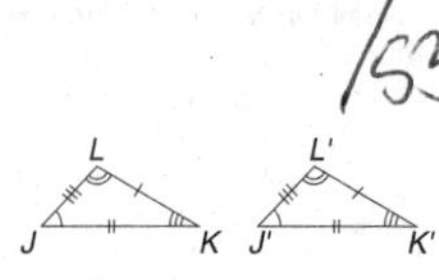
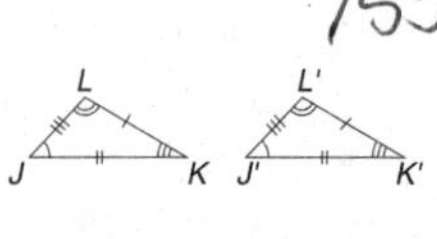

$\triangle JKL \cong \triangle J'K'L'$

Use the markings in each diagram.
Complete to write each congruence statement.

1. 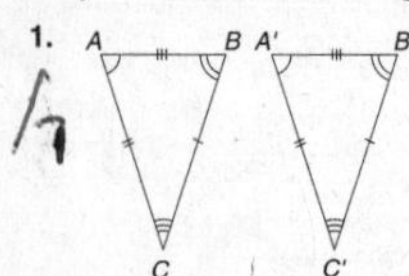

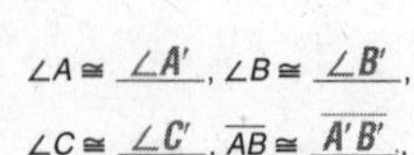

$\angle A \cong$ $\angle A'$, $\angle B \cong$ $\angle B'$,
$\angle C \cong$ $\angle C'$, $\overline{AB} \cong$ $\overline{A'B'}$,
$\overline{BC} \cong$ $\overline{B'C'}$, $\overline{AC} \cong$ $\overline{A'C'}$,
$\triangle ABC \cong$ $\triangle A'B'C'$

2. 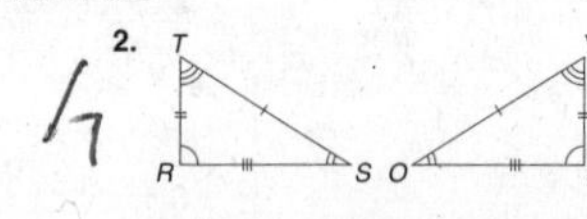

$\angle R \cong$ $\angle J$, $\angle S \cong$ $\angle O$,
$\angle T \cong$ $\angle Y$, $\overline{RS} \cong$ $\overline{JO}$,
$\overline{RT} \cong$ $\overline{JY}$, $\overline{TS} \cong$ $\overline{YO}$,
$\triangle RST \cong$ $\triangle JOY$

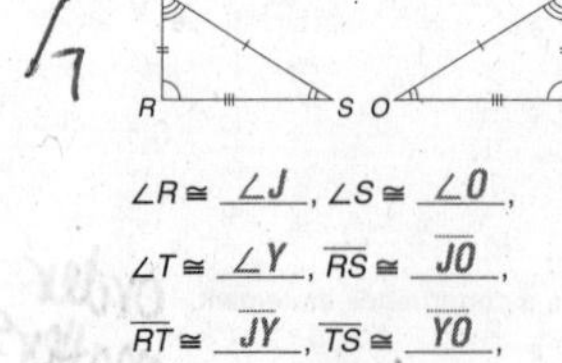
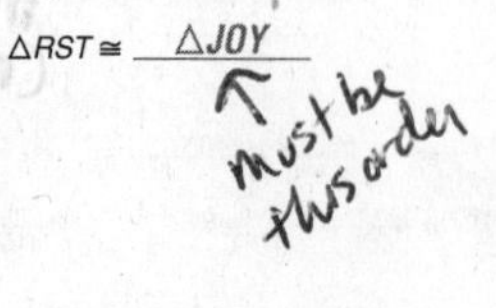

3. 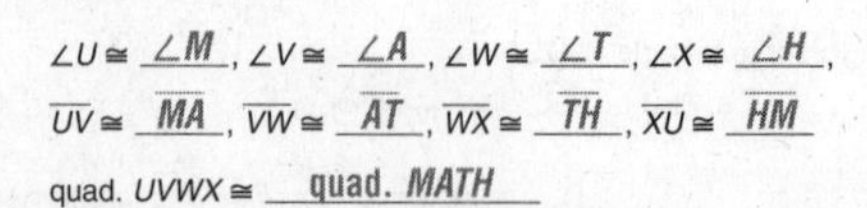

$\angle U \cong$ $\angle M$, $\angle V \cong$ $\angle A$, $\angle W \cong$ $\angle T$, $\angle X \cong$ $\angle H$,
$\overline{UV} \cong$ $\overline{MA}$, $\overline{VW} \cong$ $\overline{AT}$, $\overline{WX} \cong$ $\overline{TH}$, $\overline{XU} \cong$ $\overline{HM}$

quad. $UVWX \cong$ quad. $MATH$

LESSON 7-6
Reteach
Congruence (continued)

Congruence relations can be used to find unknown values.

$\triangle ABC \cong \triangle QPR$

$\angle A \cong \angle Q$
$3x = 90$
$\frac{3x}{3} = \frac{90}{3}$
$x = 30$

Using the congruence relationship, complete to find each unknown value.

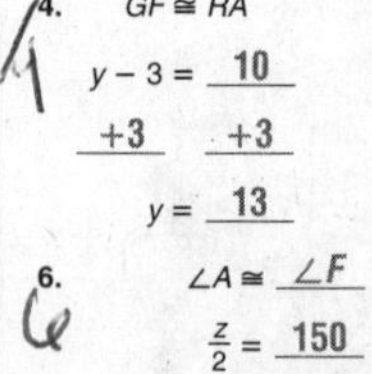
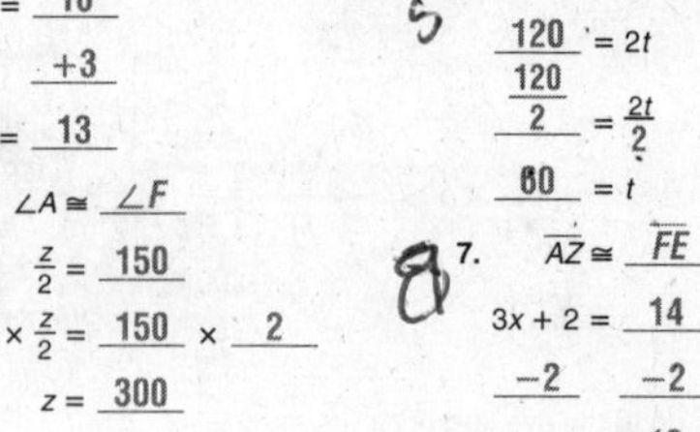

4. $\overline{GF} \cong \overline{RA}$
$y - 3 =$ 10
+3 +3
$y =$ 13

5. $\angle C \cong$ $\angle S$
120 $= 2t$
$\frac{120}{2} = \frac{2t}{2}$
60 $= t$

6. $\angle A \cong$ $\angle F$
$\frac{z}{2} =$ 150
2 $\times \frac{z}{2} =$ 150 $\times$ 2
$z =$ 300

7. $\overline{AZ} \cong$ $\overline{FE}$
$3x + 2 =$ 14
−2 −2
$3x =$ 12
$\frac{3x}{3} = \frac{12}{3}$
$x =$ 4

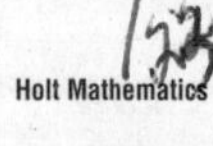

LESSON 7-6
Challenge
Cloning

In the following exercises, you will construct a triangle congruent to $\triangle ABC$ by copying three strategic parts.

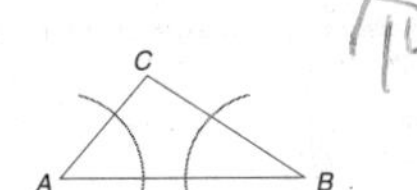

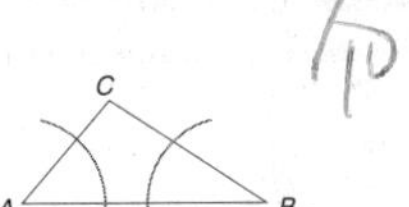

1. To construct a triangle congruent to $\triangle ABC$ by using two pairs of sides and the included angle:

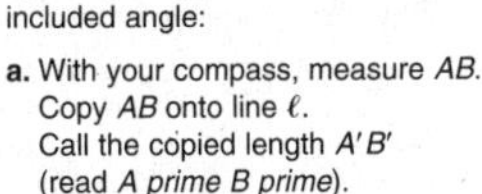

   a. With your compass, measure AB. Copy AB onto line ℓ. Call the copied length $A'B'$ (read *A prime B prime*).

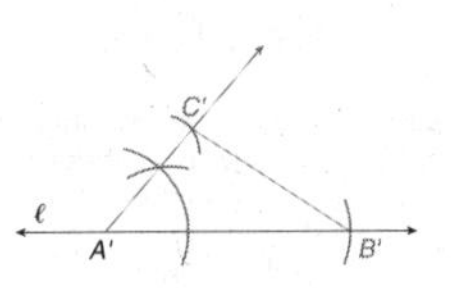

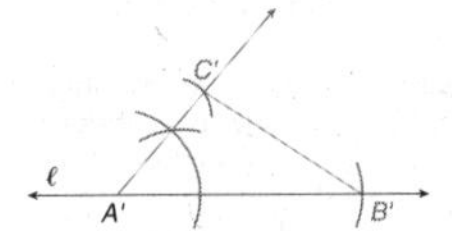

   b. With your compass, measure $\angle A$. Copy $\angle A$ at vertex A' with one side as $\overline{A'B'}$.

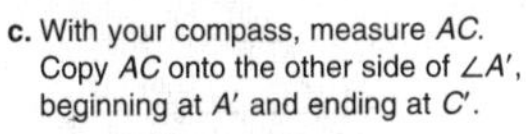

   c. With your compass, measure AC. Copy AC onto the other side of $\angle A'$, beginning at A' and ending at C'.

   d. Draw $\overline{C'B'}$. Use a ruler and protractor to verify that $\triangle A'B'C' \cong \triangle ABC$.

2. To construct a triangle congruent to $\triangle ABC$ by using two pairs of angles and the included side:

   a. With your compass, measure AB. Copy AB onto line m. Call the copied length $A''B''$ (*read A double prime B double prime*).

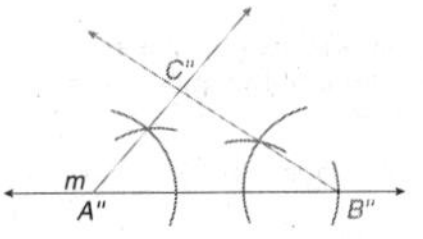

   b. With your compass, measure $\angle A$. Copy $\angle A$ at vertex A'', with one side as $\overline{A''B''}$.

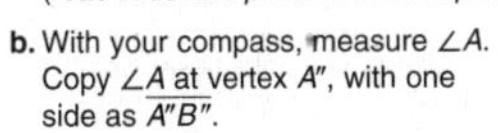

   c. With your compass, measure $\angle B$. Copy $\angle B$ at vertex B'', with one side as $\overline{A''B''}$.

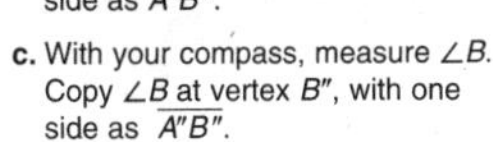

   d. Use C'' to label the point where the sides of $\angle A''$ and $\angle B''$ intersect. Use a ruler and protractor to verify that $\triangle A''B''C'' \cong \triangle ABC$.

LESSON 7-6
Problem Solving
Congruence

Use the American patchwork quilt block design called Carnival to answer the questions. Triangle $AIH \cong$ Triangle AIB, Triangle $ACJ \cong$ Triangle AGJ, Triangle $GFJ \cong$ Triangle CDJ.

1. What is the measure of $\angle IAB$?
   45°
2. What is the measure of $\overline{AH}$?
   4 inches
3. What is the measure of $\overline{AG}$?
   6 inches
4. What is the measure of $\angle JDC$?
   90°
5. What is the measure of $\overline{FG}$?
   4 inches

The sketch is part of a bridge. Trapezoid $ABEF \cong$ Trapezoid $DEBC$. Choose the letter for the best answer.

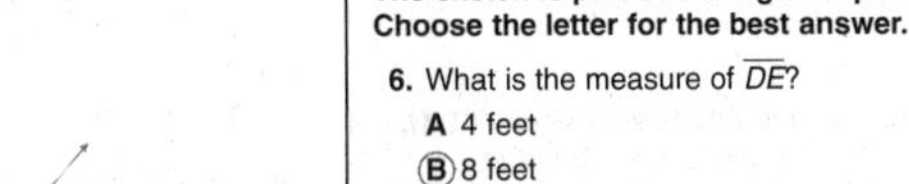

6. What is the measure of $\overline{DE}$?
   A 4 feet
   (B) 8 feet
   C 16 feet
   D Cannot be determined
7. What is the measure of $\overline{FE}$?
   F 4 feet H 8 feet
   (G) 16 feet J 24 feet
8. What is the measure of $\angle FAB$?
   A 45° C 60°
   B 90° (D) 120°
9. What is the measure of $\angle ABE$?
   F 45° H 60°
   G 90° (J) 120°
10. What is the measure of $\angle EBC$?
   A 45° (C) 60°
   B 90° D 120°
11. What is the measure of $\angle BED$?
   F 45° H 60°
   G 90° (J) 120°
12. What is the measure of $\angle BCD$?
   A 45° (C) 60°
   B 90° D 120°

LESSON 7-6
Reading Strategies
Graphic Organizer

This picture helps you understand congruence.

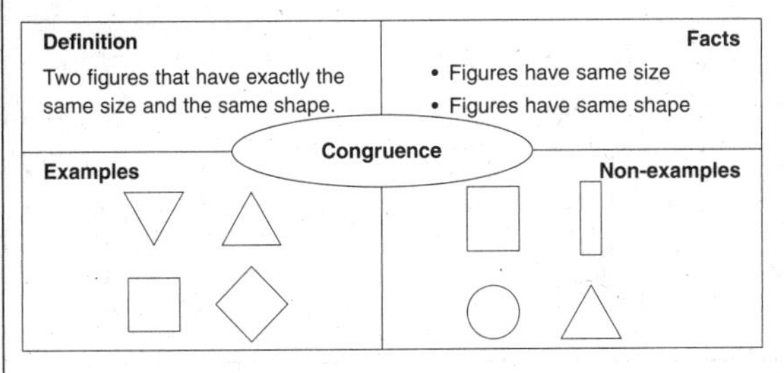

Use the picture to answer these questions.

1. How can you tell if two figures are congruent?
   They have the same size and shape.
2. Is the circle congruent to any other shape? no

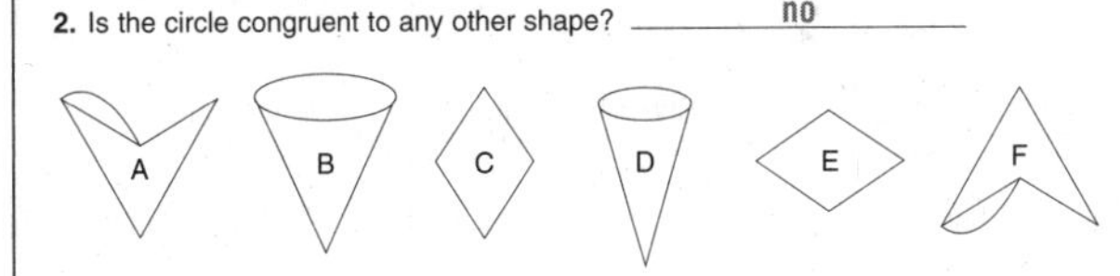

Use the chart and the figures above to answer the following questions.

3. Is figure A congruent to figure B? no
4. Is figure B congruent to figure D? no
5. Is figure C congruent to figure E? yes
6. Is figure A congruent to figure F? yes

LESSON 7-6
Puzzles, Twisters & Teasers
Equal Time!

In each problem there are congruent figures with three answers. Decide which answer is correct. Write the letter of the correct answer on the blank line which corresponds to the problem.

1. (K) $\triangle KLM \cong \triangle RQS$
   R $\angle K \cong \angle Q$
   M $\overline{KM} \cong \triangle QS$

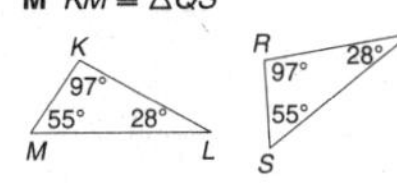

2. T $\overline{ED} \cong \overline{JF}$
   S $\angle B \cong \angle H$
   (F) pentagon $ABCDE \cong$ pentagon $JIHGF$

3. (E) $\overline{AB} \cong \overline{FE}$
   G $\triangle ACB \cong \triangle DEF$
   W $\triangle CAB \cong \triangle EFD$

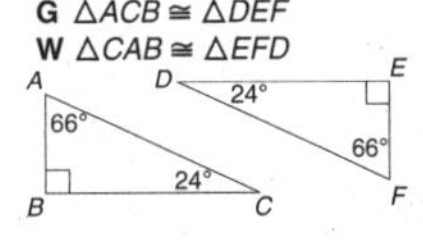

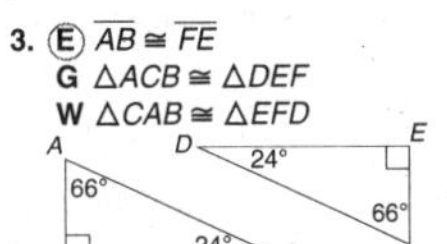

4. K quadrilateral $LMNO \cong$ quadrilateral $QRST$
   (L) $\angle N \cong \angle T$
   S $\angle L \cong \angle T$

5. (C) $\triangle CAB \cong \triangle JKL$
   D $\angle A \cong \angle L$
   N $ACB \cong KLJ$

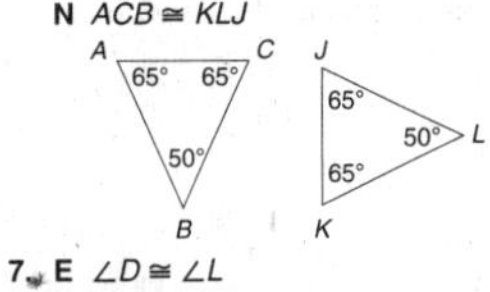

6. I quadrilateral $PQSR \cong$ quadrilateral $TUVW$
   (A) $\angle P \cong \angle U$
   O $\overline{RP} \cong \overline{TW}$

7. E $\angle D \cong \angle L$
   A $\overline{AD} \cong \overline{ML}$
   (O) $\overline{AB} \cong \overline{NM}$

What kind of face has hands but no eyes, nose, or mouth?

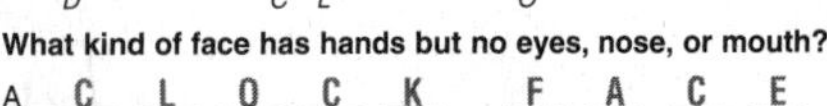

A C L O C K F A C E
5 4 7 5 1 2 6 5 3

LESSON 7-7
Practice A
Transformations

Identify each as a translation, rotation, reflection, or none of these.

1.
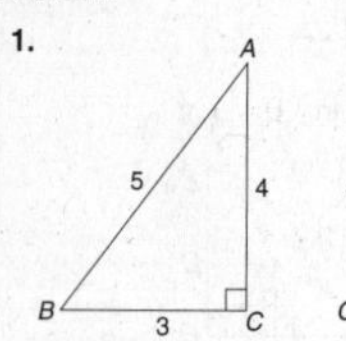
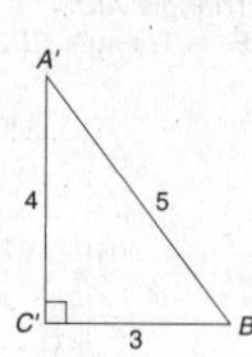

reflection

2.
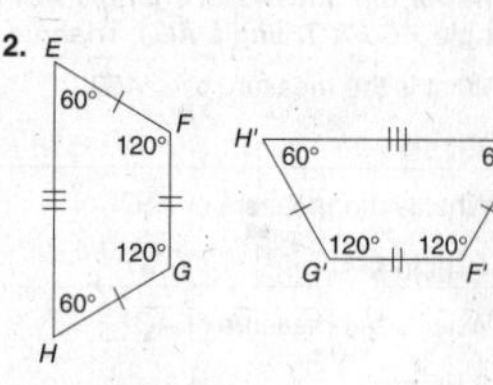

rotation

3.
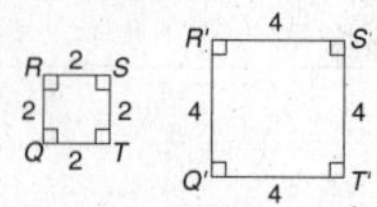

none of these

4. A 10 B A' 10 B' 6 6 6 6 D 10 C D' 10 C'

translation

5. Draw the image of the triangle ABC with vertices (−2, 2), (2, 4), and (2, 2) after a translation 5 units down.

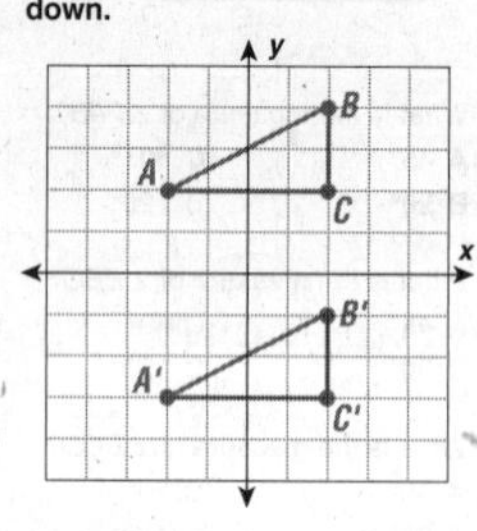

6. Triangle XYZ has vertices X(3, 4), Y(4, 1), and Z(1, 1). Find the coordinates of the image of point Z after a reflection across the y-axis.

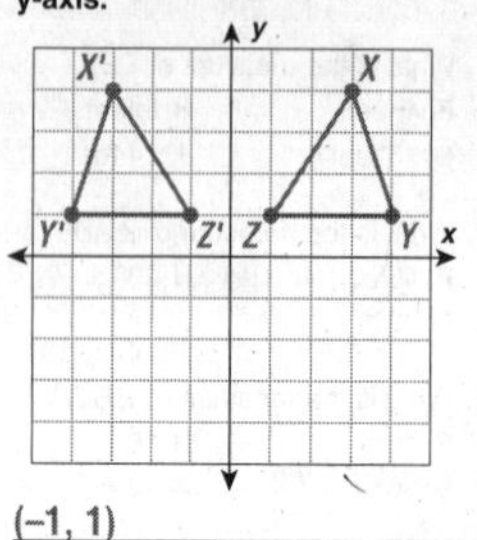

(−1, 1)

16

LESSON 7-7
Practice B
Transformations

Identify each as a translation, rotation, reflection, or none of these.

1.
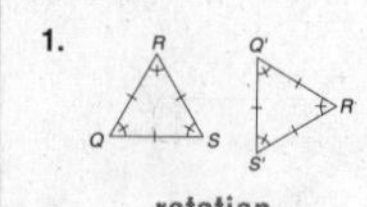

rotation

2.
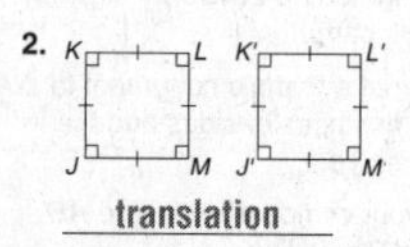

translation

Draw the image of the rectangle ABCD with vertices (−2, 1), (−1, 3), and (3, 3), (2, 1) after each transformation.

3. translation 3 units down

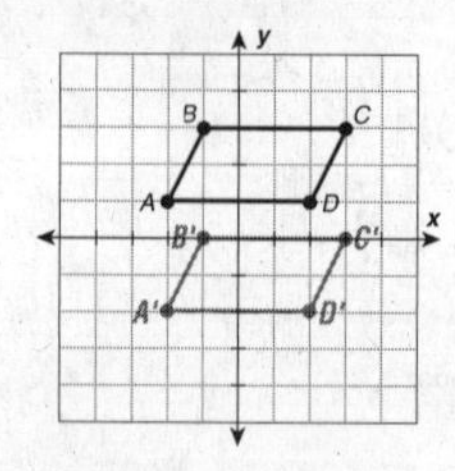

4. 180° rotation around (0, 0)

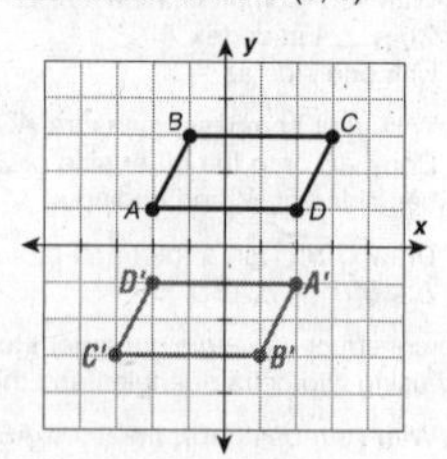

Triangle ABC has vertices A(−3, 1), B(2, 4), and C(3, 1). Find the coordinates of the image of each point after each transformation.

5. reflection across the x-axis, point B

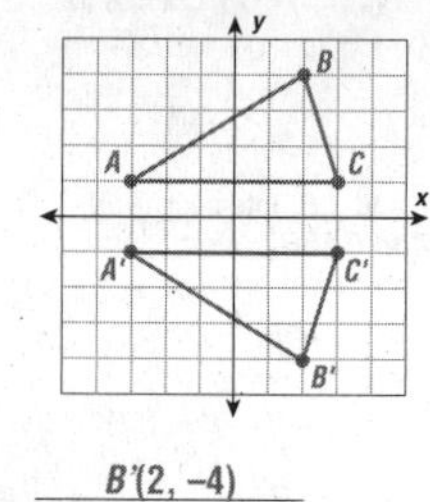

B'(2, −4)

6. translation 6 units down, point A

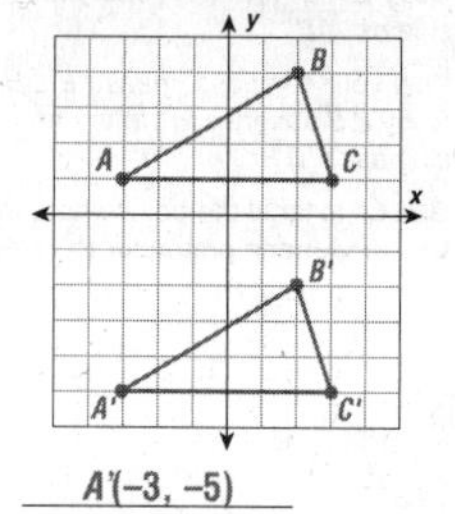

A'(−3, −5)

LESSON 7-7
Practice C
Transformations

Give the coordinates of each point after a reflection across the x-axis.

1. (−2, 3) — (−2, −3)
2. (−4, −1) — (−4, 1)
3. (5, 2) — (5, −2)
4. (6, −3) — (6, 3)

Give the coordinates of each point after a reflection across the y-axis.

5. (−1, −5) — (1, −5)
6. (3, 2) — (−3, 2)
7. (−4, 6) — (4, 6)
8. (7, −2) — (−7, −2)

Give the coordinates of each point after a 180° rotation around (0, 0).

9. (4, −6) — (−4, 6)
10. (−5, 3) — (5, −3)
11. (1, 2) — (−1, −2)
12. (−3, −2) — (3, 2)

Perform the given transformation.

13. Reflect across line m.

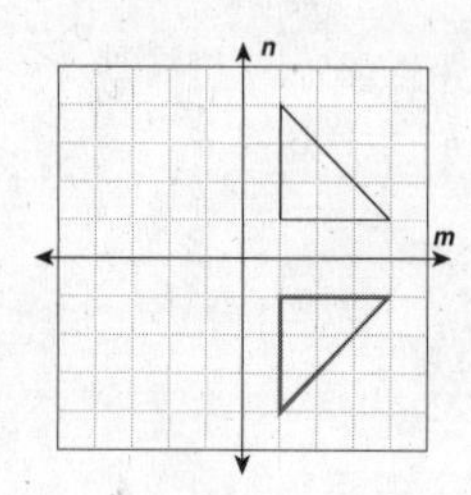

14. Rotate clockwise 180° around (0, 0).

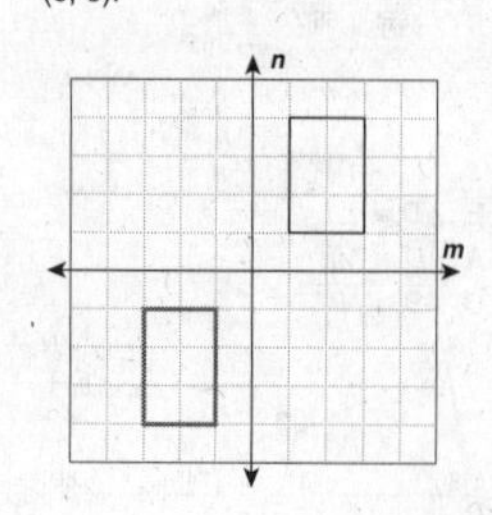

14

LESSON 7-7
Reteach
Transformations

Reflection	Rotation	Translation
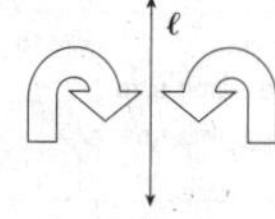	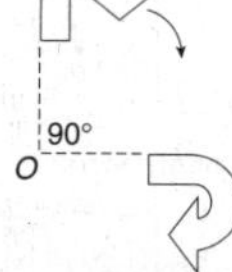 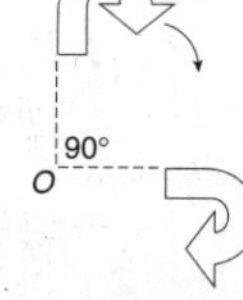	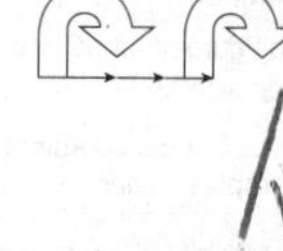
a *mirror image* (a *flip*) The figure is reflected over line ℓ.	a *turning* The figure is rotated 90° clockwise about point O.	a *slide* The figure is translated 3 units right and 1 unit up.

12

Complete to identify each type of transformation.

1.
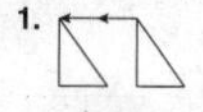

Slide the figure 2 units left.

Transformation: translation

2.
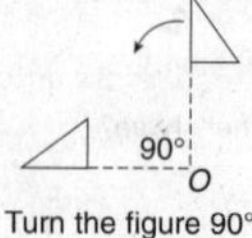

Turn the figure 90° counterclockwise.

Transformation: rotation.

3. Flip the figure over line m.

Transformation: reflection

Identify each as a translation, rotation, or reflection. ???

4. A k A' B C C' B' — reflection

5.

rotation.

6. A A' B C B' C' — translation

16

LESSON 7-7 Reteach
Transformations (continued)

When reflecting a point about a horizontal or vertical line, only one of the coordinates changes.

reflection across *y*-axis
x-coordinate goes to its opposite

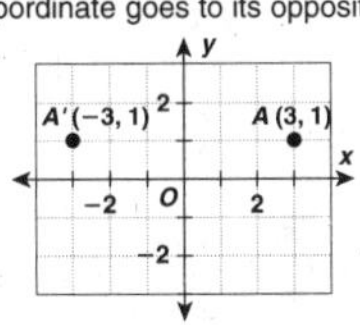

reflection across *x*-axis
y-coordinate goes to its opposite

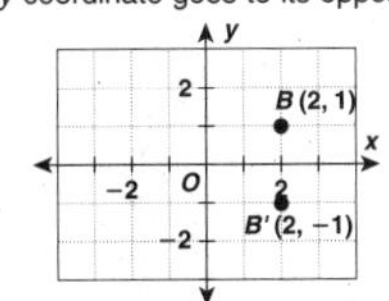

When translating a point, add the indicated number of units to each coordinate.

For a translation left or right, add units to the x-coordinate. For a translation up or down, add units to the y-coordinate.

P(1, 4) is translated 3 units down.

P(1, 4) → P′(1, 4 + (−3)), or P′(1, 1)

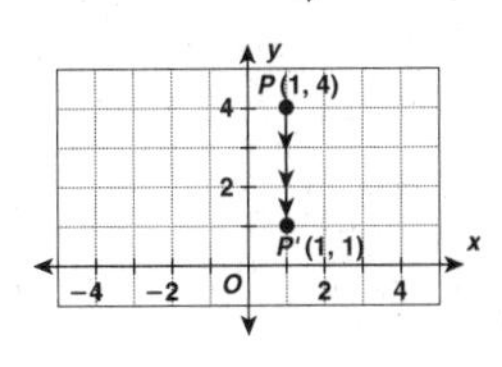

Draw and label the image after the reflection.

7. *P*(−1, 2) over the *y*-axis

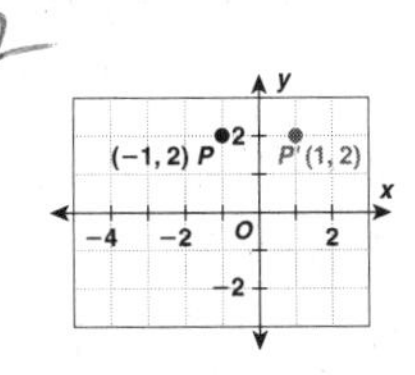

Draw and label the image after the translation.

8. Translate *A*(−3, 5) 4 units to the right.

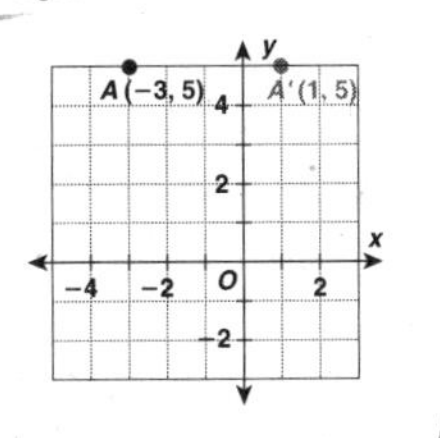

LESSON 7-7 Challenge
One Trip Instead of Two

1. Consider △*ABC* with vertices at *A*(4, 1), *B*(1, 2), and *C*(3, 5).

a. Draw △*A′B′C′*, the image of △*ABC* after a reflection in the line *x* = 1.

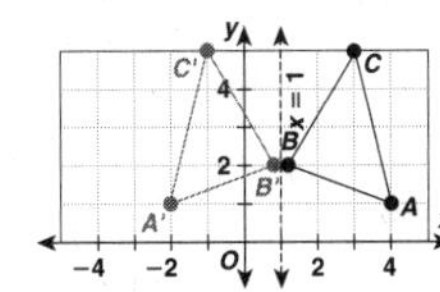

b. Draw △*A″B″C″*, the image of △*A′B′C′* after a reflection in the *y*-axis.

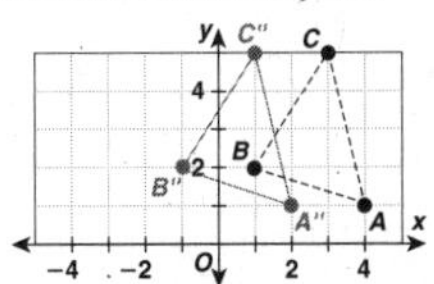

c. Describe a single transformation that takes △*ABC* to the image △*A″B″C″*.

translation of 2 units to the left

d. How are the lines of reflection related?

parallel lines

2. Consider △*PQR* with vertices at *P*(1, 2), *Q*(3, 4), and *R*(5, 3).

a. Draw △*P′Q′R′*, the image of △*PQR* after a reflection in the line *y* = 1.

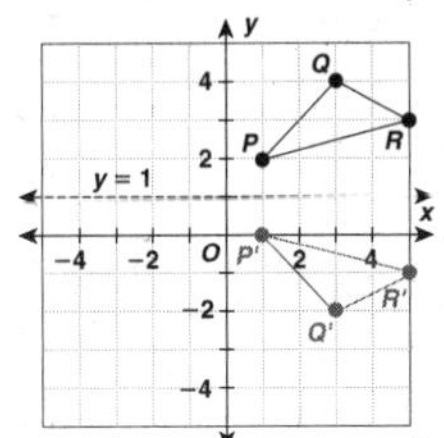

b. Draw △*P″Q″R″*, the image of △*P′Q′R′* after a reflection in the *x*-axis.

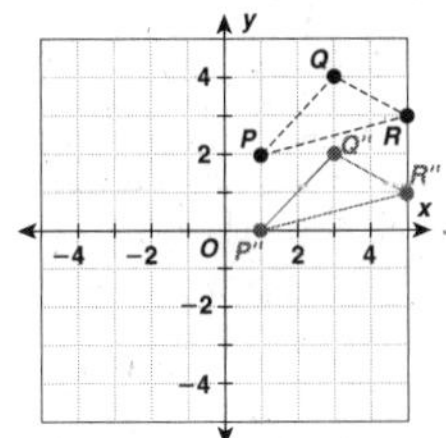

c. Describe a single transformation that takes △*PQR* to the image △*P″Q″R″*.

translation of 2 units down

d. How are the lines of reflection related?

parallel lines

LESSON 7-7 Problem Solving
Transformations

Parallelogram ABCD has vertices *A*(−3, 1), *B*(−2, 4), *C*(3, 4), and *D*(2, 1). Refer to the parallelogram to write the correct answer.

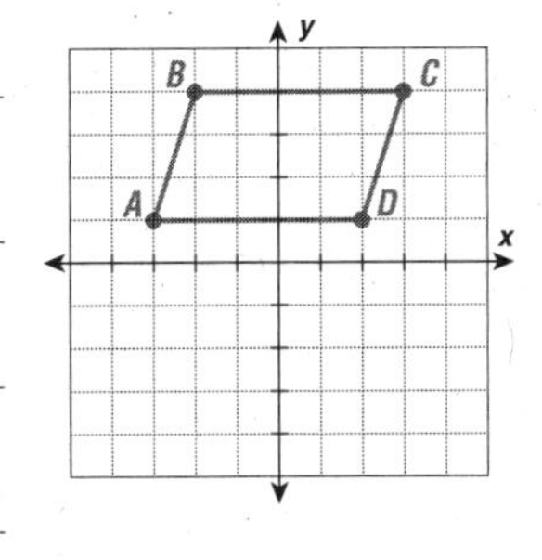

1. What are the coordinates of point A after a reflection across the x-axis?

(−3, −1)

2. What are the coordinates of point B after a reflection across the y-axis?

(2, 4)

3. What are the coordinates of point C after a translation 2 units down?

(3, 2)

4. What are the coordinates of point D after a 180° rotation around (0, 0)?

(−2, −1)

Identify each as a translation, rotation, reflection or none of these.

5. 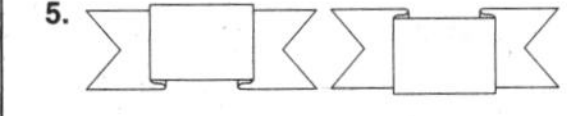

A translation
B reflection
Ⓒ rotation
D none of these

6.

F translation
Ⓖ reflection
H rotation
J none of these

7.

Ⓐ translation
B reflection
C rotation
D none of these

8. 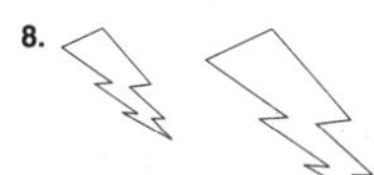

F translation
G reflection
H rotation
Ⓙ none of these

LESSON 7-7 Reading Strategies
Focus on Vocabulary

Transformation is a change in the position or size of a figure. The act of transforming the figure does not change its shape.

Here are three kinds of transformations that do not change a figure's size or shape.

- **Translation** → **Slide** a figure to a new position.

Slide

- **Rotation** → **Turn** a figure around a point to a new position.

Rotate

- **Reflection** → **Flip** a figure over a line for a mirror image.

Flip

Use *translation, rotation,* or *reflection* to name the transformation shown in each picture.

1. reflection

2. rotation

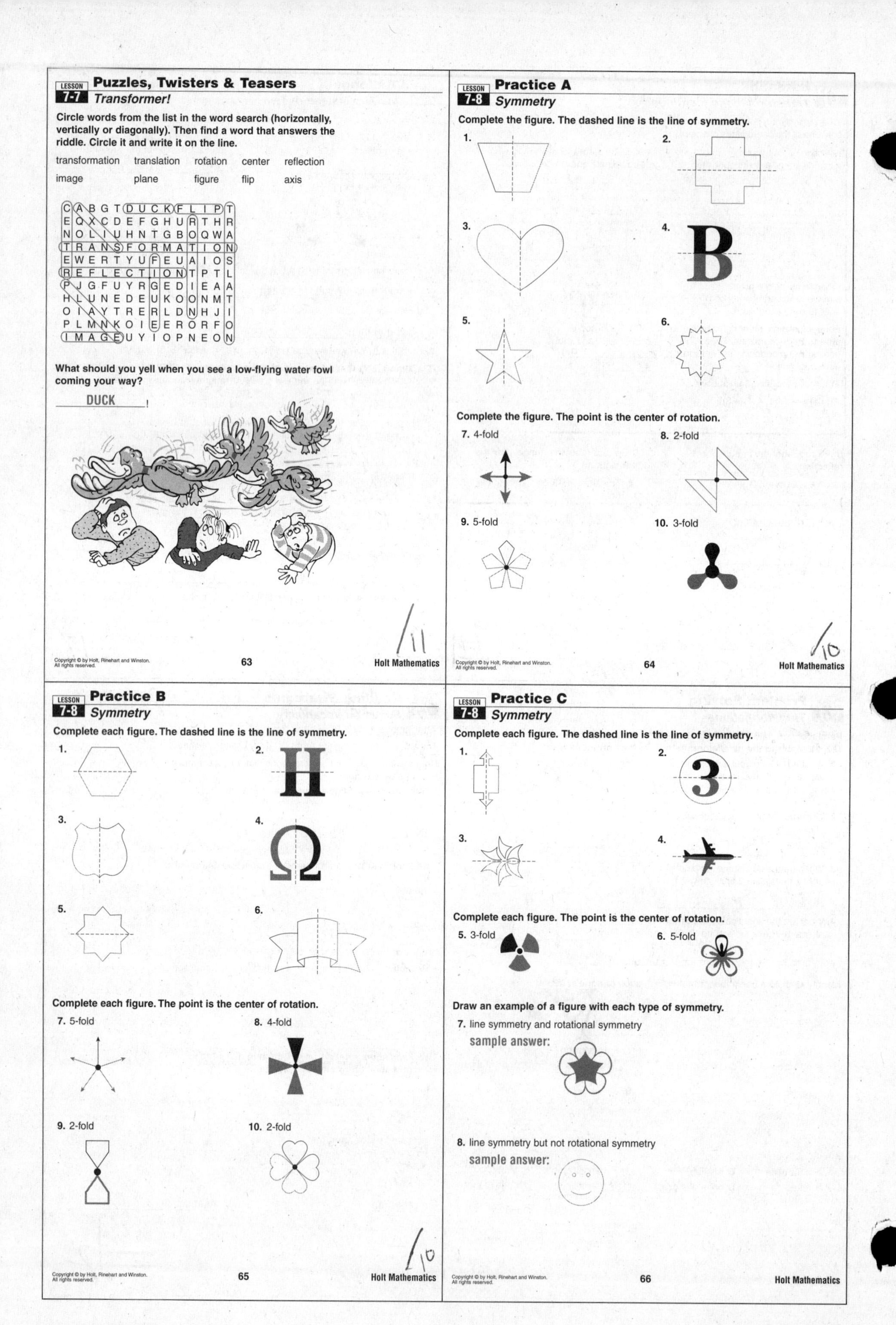

LESSON 7-7 **Puzzles, Twisters & Teasers**
Transformer!

Circle words from the list in the word search (horizontally, vertically or diagonally). Then find a word that answers the riddle. Circle it and write it on the line.

transformation translation rotation center reflection
image plane figure flip axis

C A B G T D U C K F L I P T
E Q X C D E F G H U R T H R
N O L I U H N T G B O Q W A
T R A N S F O R M A T I O N
E W E R T Y U F E U A I O S
R E F L E C T I O N T P T L
P J G F U Y R G E D I E A A
H L U N E D E U K O O N M T
O I A Y T R E R L D N H J I
P L M N K O I E E R O R F O
I M A G E U Y I O P N E O N

What should you yell when you see a low-flying water fowl coming your way?

DUCK!

/11

LESSON 7-8 **Practice A**
Symmetry

Complete the figure. The dashed line is the line of symmetry.

1. 2. 3. 4. 5. 6.

Complete the figure. The point is the center of rotation.

7. 4-fold
8. 2-fold
9. 5-fold
10. 3-fold

/10

LESSON 7-8 **Practice B**
Symmetry

Complete each figure. The dashed line is the line of symmetry.

1. 2. 3. 4. 5. 6.

Complete each figure. The point is the center of rotation.

7. 5-fold
8. 4-fold
9. 2-fold
10. 2-fold

/10

LESSON 7-8 **Practice C**
Symmetry

Complete each figure. The dashed line is the line of symmetry.

1. 2. 3. 4.

Complete each figure. The point is the center of rotation.

5. 3-fold
6. 5-fold

Draw an example of a figure with each type of symmetry.

7. line symmetry and rotational symmetry
sample answer:

8. line symmetry but not rotational symmetry
sample answer:

LESSON 7-8
Reteach
Symmetry

Vertical Line Symmetry | Horizontal Line Symmetry | Diagonal Line Symmetry

Draw all the lines of symmetry in each figure.

1. 1
2. 4
3. 5

Rotational symmetry is when you can turn an object so that it looks exactly the same. The number of positions in which it looks exactly the same gives you its order of symmetry.

A — Position 1 | A — Position 2 | A — Position 3

Follow vertex *A* to see how the figure is turned.
Just twice, the figure looks exactly the same, in Positions 1 and 3.
So, this figure has 2-fold rotational symmetry.

Tell the order of rotational symmetry.

4. 2-fold
5. 3-fold
6. 4-fold

6/6

LESSON 7-8
Challenge
Inside, Outside

Materials: sheet of paper (8.5 in. by 11 in.), scissors, tape

1. **a.** Cut a strip of paper 2 in. by 11 in. Tape the ends together to form a loop.
   b. Use a pencil to create a center line around the outside of the loop. Cut the loop along its center line. How many loops? 2

2. **a.** Cut a strip of paper 2 in. by 11 in. Before taping the ends together, give the strip a half-twist by turning one end over 180°.
   b. Use a pencil to create a center line around the loop. Continue drawing until you reach your starting point. Describe the result. How many surfaces (sides) does this strip have?
   The line appears on both sides of the strip; one surface.
   c. Cut the loop along its center line. How many loops do you have?
   1 loop

This unusual surface is called a **Mobius Strip,** named after A. F. Mobius, a 19th century German mathematician and astronomer, who pioneered *topology* (how geometric figures act when distorted by such things as twists and stretches).

3. Prepare another Mobius Strip, a loop with a half-twist. Cut this strip parallel to one edge about one-third of the way from the edge. Continue cutting until you get back to your starting point. Describe the result.
   2 loops of unequal lengths linked together
4. Prepare a loop with a full twist. Cut this strip parallel to one edge about one-third of the way from the edge. Continue cutting until you get back to your starting point. Describe the result.
   2 loops of same length linked together; one loop is twice as wide as the other
5. Prepare a loop with three half-twists. Cut this strip parallel to one edge about one-half of the way from the edge. Continue cutting until you get back to your starting point. Describe the result.
   1 loop with a knot

1/6

LESSON 7-8
Problem Solving
Symmetry

Complete the figure. A dashed line is a line of symmetry and a point is a center of rotation.

1.
2.
3. 2 fold symmetry
4. 4 fold symmetry

Use the flag of Switzerland to answer the questions.

5. Which of the following would NOT be a line of symmetry?
   A $\overline{HD}$ C $\overline{AE}$
   B $\overline{BF}$ (D) $\overline{HB}$
6. How many lines of symmetry does the flag have?
   F 2 (H) 4
   G 6 J 8
7. How many folds of rotational symmetry does the flag have?
   A 0 C 2
   (B) 4 D 8
8. Which lists all lines of symmetry of the flag?
   F $\overline{AE}$, $\overline{GC}$
   G $\overline{HD}$, $\overline{BF}$
   (H) $\overline{HD}$, $\overline{BF}$, $\overline{AE}$, $\overline{GC}$
   J $\overline{HB}$, $\overline{DF}$, $\overline{AE}$, $\overline{GC}$
9. Which describes the center of rotation?
   (A) intersection of $\overline{BF}$ and $\overline{HD}$
   B intersection of $\overline{AE}$ and $\overline{HB}$
   C *A*
   D There is no center of rotation

LESSON 7-8
Math: Reading and Writing in the Content Area
Reading a Diagram

The star below turns around a center point. If a figure rotates onto itself before one complete rotation, the figure has **rotational symmetry.**

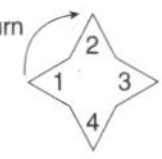
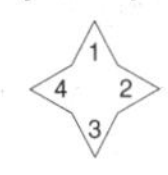

The star matches up with the original figure before it has made one complete turn. It has rotational symmetry.

This right triangle does not have rotational symmetry. The rotated image matches up with the original triangle only after it has made one complete turn.

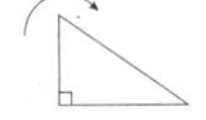
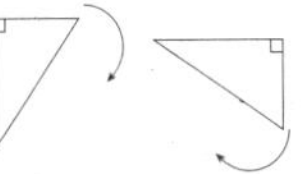
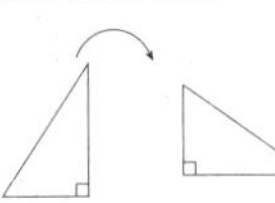

Do these figures have rotational symmetry? Trace each figure and rotate it to find out. Write *yes* or *no*.

1. yes
2. no

A figure can be reflected over a line so that the two parts are congruent. That figure has **line symmetry.**

Line Symmetry | No Line Symmetry

Do these figures have line symmetry? Trace each figure. Draw lines of symmetry if possible. Write *yes* or *no*.

3. yes
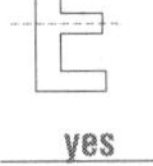
4. no

4/4

LESSON 7-8
Puzzles, Twisters & Teasers
Concentrating on Symmetrical Figures

Imagine a game of concentration. Each box represents a card in the game. Choose two cards with figures that would have symmetry if they were put together. Cross them out. Rearrange the bold letters of the unmatched cards to solve the riddle.

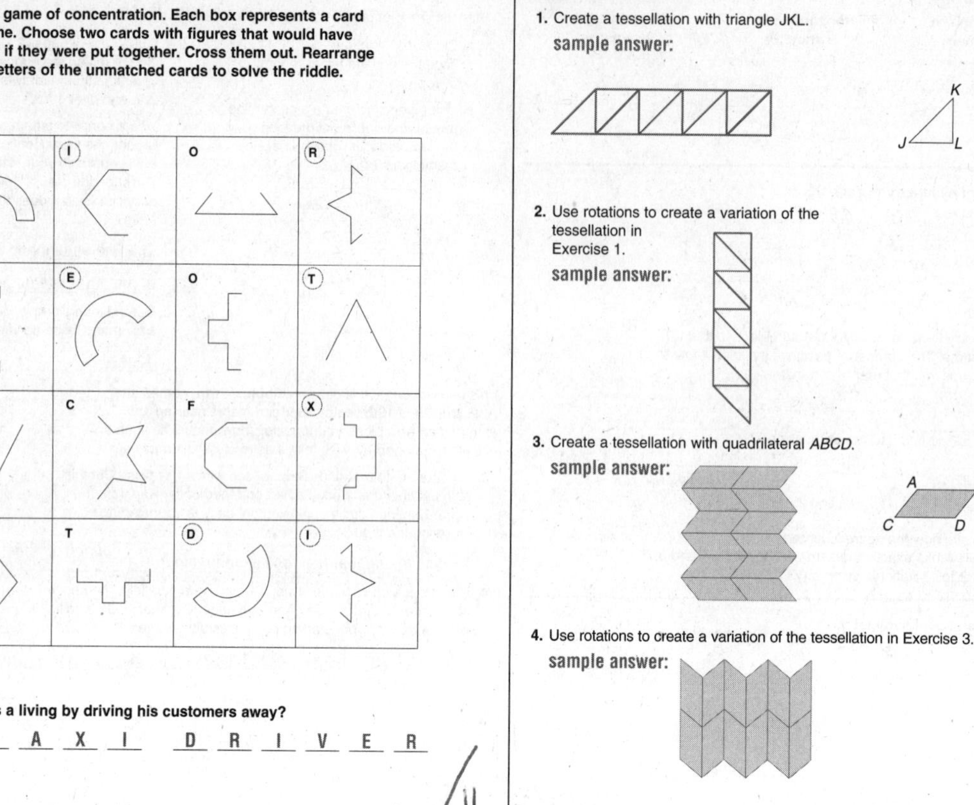

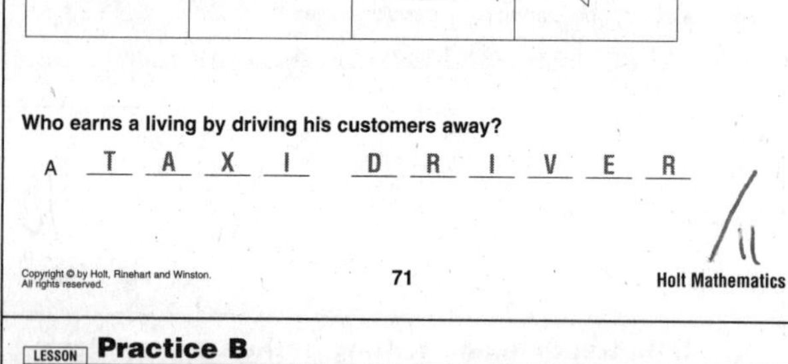

Who earns a living by driving his customers away?

A T A X I D R I V E R

LESSON 7-9
Practice A
Tessellations

1. Create a tessellation with triangle JKL.
   sample answer:

2. Use rotations to create a variation of the tessellation in Exercise 1.
   sample answer:

3. Create a tessellation with quadrilateral *ABCD*.
   sample answer:

4. Use rotations to create a variation of the tessellation in Exercise 3.
   sample answer:

LESSON 7-9
Practice B
Tessellations

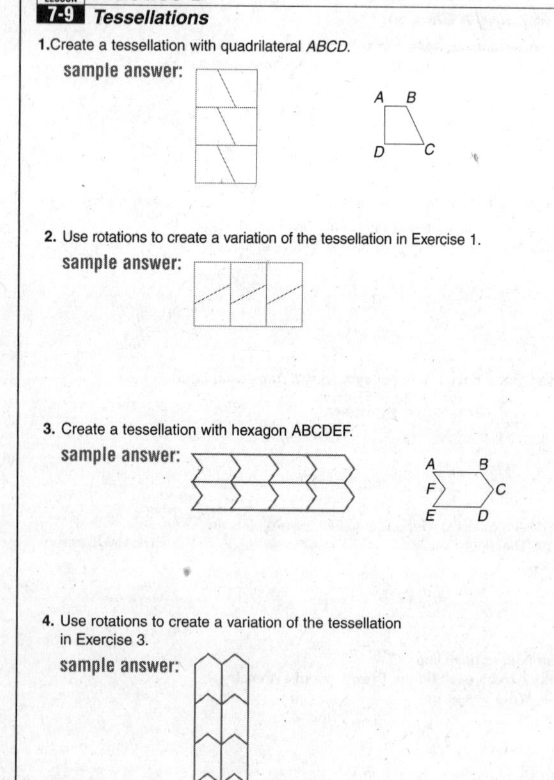

1. Create a tessellation with quadrilateral *ABCD*.
   sample answer:

2. Use rotations to create a variation of the tessellation in Exercise 1.
   sample answer:

3. Create a tessellation with hexagon ABCDEF.
   sample answer:

4. Use rotations to create a variation of the tessellation in Exercise 3.
   sample answer:

LESSON 7-9
Practice C
Tessellations

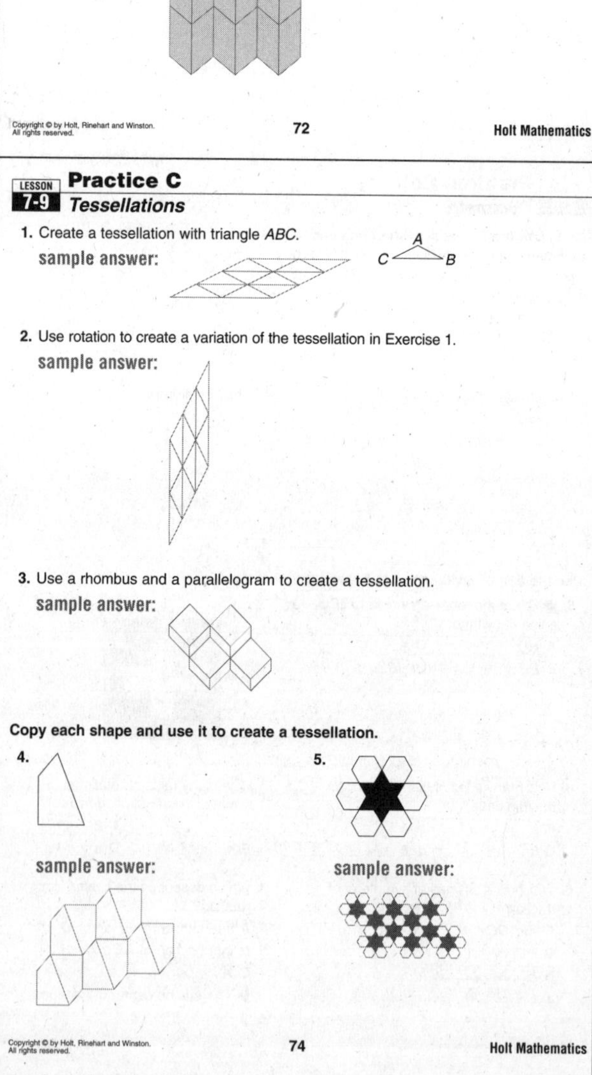

1. Create a tessellation with triangle *ABC*.
   sample answer:

2. Use rotation to create a variation of the tessellation in Exercise 1.
   sample answer:

3. Use a rhombus and a parallelogram to create a tessellation.
   sample answer:

Copy each shape and use it to create a tessellation.

4. sample answer:

5. sample answer:

LESSON **7-9**

Reteach
Tessellations

A **tessellation** is a repeating pattern of shapes that completely covers a plane with no gaps or overlaps.

Not all plane shapes can be used to make a tessellation.

To make a tessellation of a figure, fit copies of the shape together.
Remember, there can be no overlaps or gaps when you fit the shapes together.

You can make a tessellation using this hexagon.

The figures fit together with no gaps or overlaps.

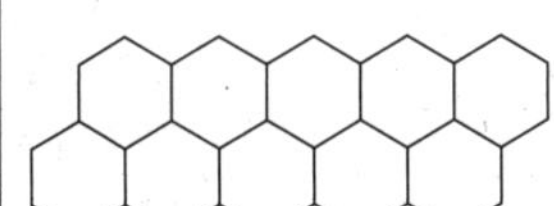

You cannot make a tessellation using this pentagon.

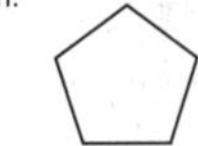

The figures do not all fit together. There is a gap.

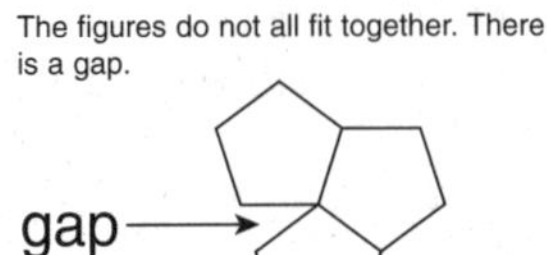

Can you make a tessellation using each shape?
Write yes or no.

1. 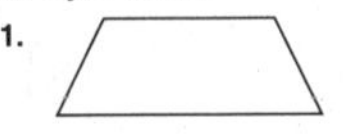

yes

2. 

no

Fit triangles shaped like this together to make a tessellation.

3. 

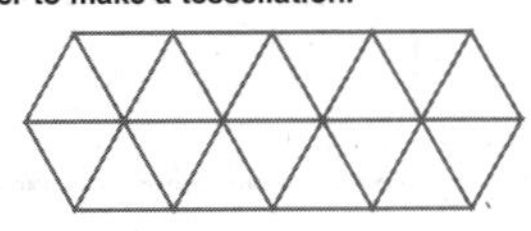

Sample answer:

LESSON **7-9**

Reteach
Tessellations (continued)

After making a **tessellation** of a figure, you can use a rotation to create a variation of the tessellation.

Use the figure shown at the right to create a tessellation.

Tessellation

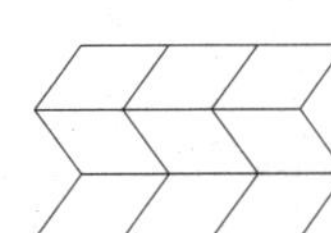

Variation
Rotate the tessellation 90° clockwise to vary the tessellation.

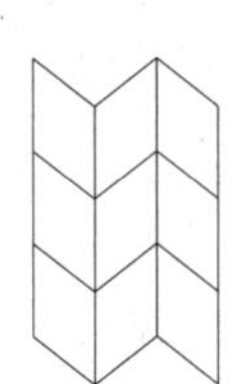

4. Make a tessellation using the shape?

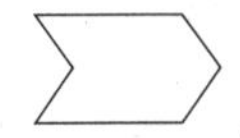

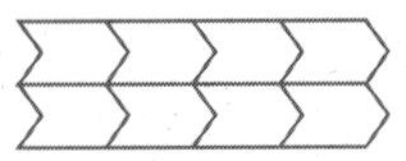

Sample answer:

5. Rotate your tessellation 90° clockwise to vary the tessellation.

Sample answer:

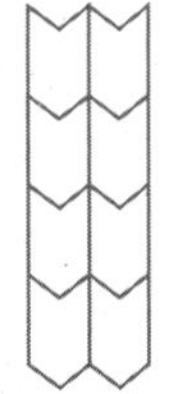

LESSON **7-9**

Challenge
Shapely

A **fractal** is a shape that repeats itself in a pattern.
As more stages are generated, the shape becomes more complex.

Starting with a + sign and generating the pattern by adding a half-size + sign in each of the four corners, the first two stages of a fractal are shown.

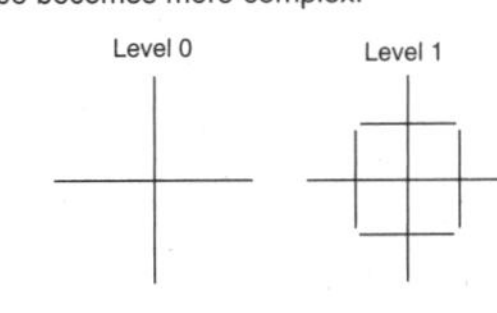

1. Generate the next two stages of the fractal.

Level 2

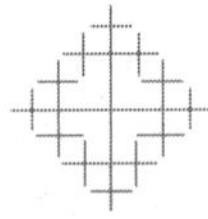

Level 3

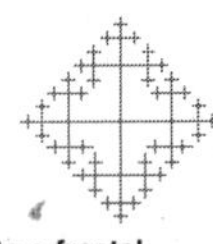

Use paper folding to generate a fractal.
Level 0: Begin with a long strip of paper.

2. Level 1: Fold the strip right end to left; crease it. Open the strip so that it forms a right angle. Stand the strip on edge on your desk. Look down and sketch this view.

3. Level 2: Start at Level 0. Do Level 1 (the pattern) twice. Open the strip so that it forms right angles. Stand the strip on edge on your desk. Look down and sketch this view. Comment on the result.

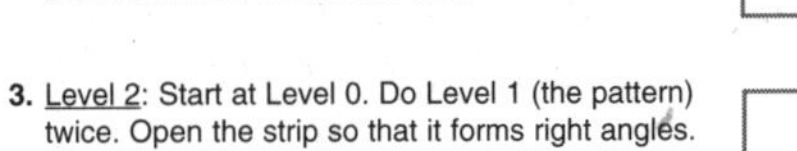

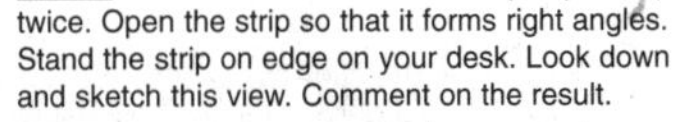

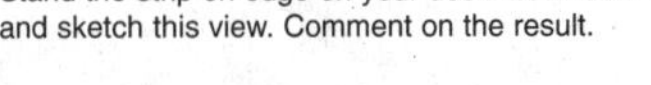

Each side of the right ∠ has been replaced by a right ∠, with sides half the previous.

4. Level 3: Start at Level 0. Do the pattern three times.

5. Level 4: Start at Level 0. Do the pattern four times.

LESSON **7-9**

Problem Solving
Tessellations

Create a tessellation using the given figure.

1.

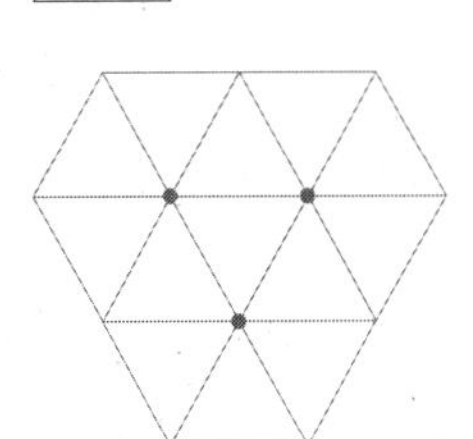

2. 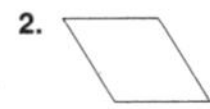

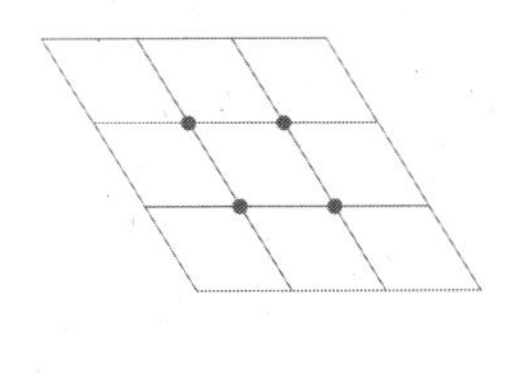

Choose the letter for the best answer.

3. Which figure will NOT make a tessellation?

A

B 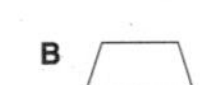

C 

Ⓓ 

4. Which nonregular polygon can always be used to tile a floor?
   F pentagon
   Ⓖ triangle
   H octagon
   J hexagon

5. For a combination of regular polygons to tessellate, the angles that meet at each vertex must add to what?
   A 90°
   B 180°
   Ⓒ 360°
   D 720°

LESSON 7-9

Reading Strategies

Draw Conclusions

When figures are arranged in a repeating pattern that covers a surface with no gaps or overlaps, the pattern is called a **tessellation.** This tessellation is made with equilateral triangles.

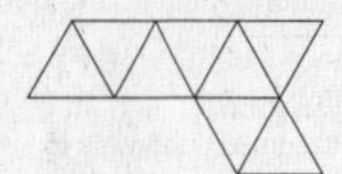

Repeating regular pentagons do not cover a surface completely. There are gaps between the pentagons, so this figure does not tessellate.

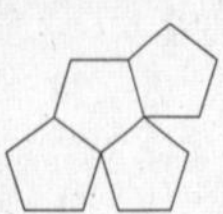

Trace and cut out several of the figures below. Test to see if the figures fit together to form a tessellation. Write *yes* or *no*.

1.

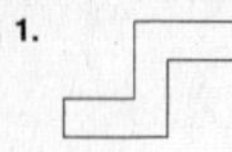

yes

2.

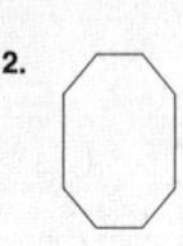

no

3.

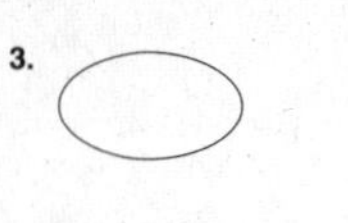

no

4.

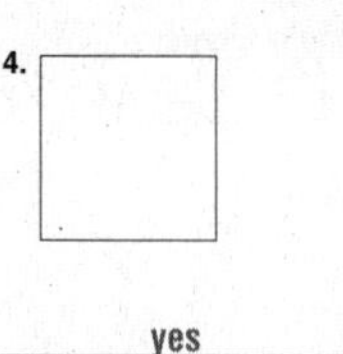

yes

LESSON 7-9

Puzzles, Twisters & Teasers

A Puzzling Pattern!

Solve the crossword puzzle.

Across

1. A ______ is a repeating pattern of a plane figure that covers a plane with no gaps or overlaps.

5. A quadrilateral with 4 equal sides and 4 equal angles is a ______.

6. To move a figure around a central point is to ______.

7. A 6-sided polygon is a ______.

8. The angle measures of a ______ add to 360°.

9. A repeating ______ is a design created by using the same figure in sequence over and over.

Down

2. In a ______ tessellation, two or more regular polygons are repeated to fill the plane.

3. An 8-sided polygon is an ______.

4. When the same figure is copied over and over again, it is ______.

6. In a ______ polygon, all sides are equal.

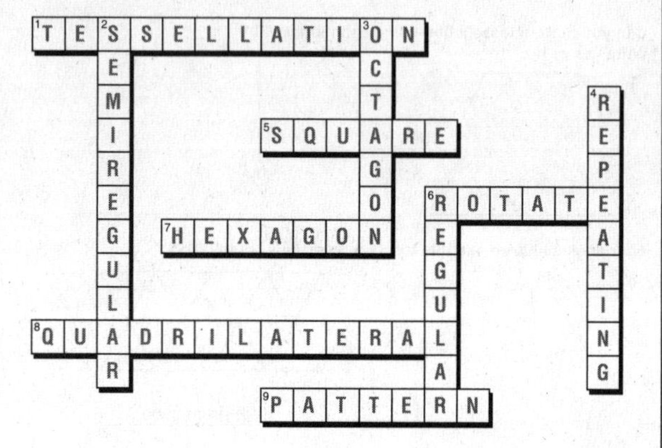